AF316603

MASTERING PROJECT MANAGEMENT

Book 1

Planning For Performance

Competency-Based Guidebook Focused on Technical Project Management

First Edition 2023

ISBN: 979-8-218-13480-8

Centrestar Learning
State College, Pennsylvania, USA
www.centrestar.com

MASTERING PROJECT MANAGEMENT

Book 1

Planning For Performance

A Competency-Based Guidebook Focused on
Technical Project Management

Preface

Mastering Project Management is a three-part book series designed for individuals who manage or anticipate managing workplace projects such as technical specialists, apprentices, frontline leaders, engineers, managers, and project stakeholders.

Book 1, ***Planning for Performance,*** focuses on the technical side of the traditional project management life cycle and covers the tools and techniques most widely used to manage workplace projects. The two companion books ***Results Through People*** and **Controlling Time, Money, and Risk,** are focused on the leadership and business management sides of project management, respectively.

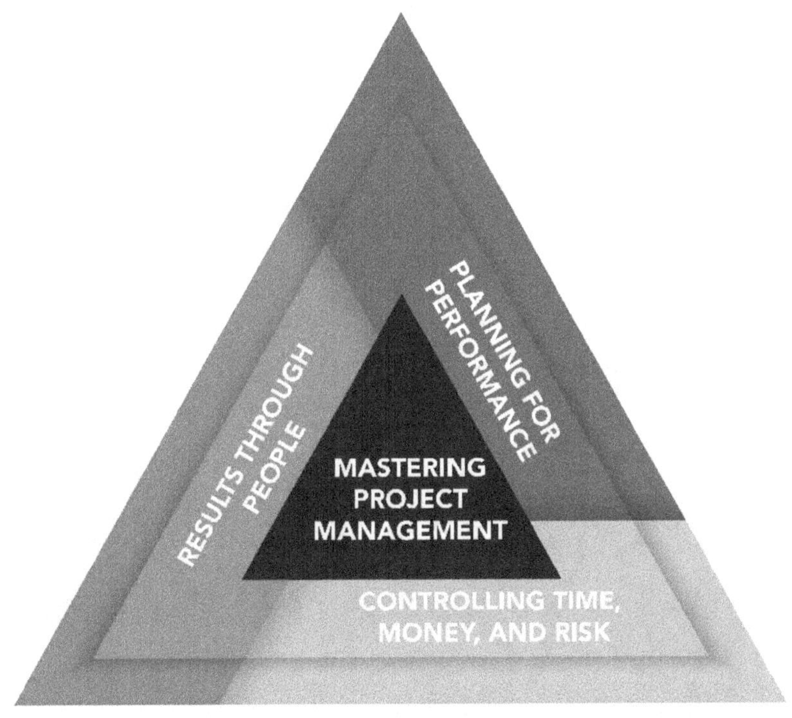

The series grew out of my being an engineer, project manager, and entrepreneur with broader leadership interests. Early in my career I experienced and observed the impact that other project managers and leaders had on organizations, both positive and negative. As a result, I pursued an MBA. I knew that I needed to learn both the "numbers" side of business to justify projects that I thought we should pursue, as well as the "soft skill" and leadership side to better deal with people problems. As I progressed through my MBA curriculum, I gained an appreciation for far more; later in my career I ultimately pursued a Ph.D. in Workforce Education and Development with an emphasis on Organization Development.

As a business owner, I learned that without customers and loyal employees you have nothing. This gave me a new appreciation for the value and impact of effective marketing. Along with this experience came the insight that to retain good employees the golden rule of "treating people the way you want to be treated" is not just a saying, but a rule for us to live by.

Anyone at the operational or senior levels of an organization knows the value and necessity of taking a "bottoms-up" approach to strategic planning. This translates to asking the people doing the work managing your projects, and the people who buy and use your products or services, for their input.

No matter what leadership level you achieve, one thing remains true; you either get better or you get worse. This is true for individuals as well as organizations. This book is based on sound research, years of experience in a variety of industries, and the know-how gained in managing projects and teaching others how to be successful leaders. Effective project managers and other leaders achieve success through a combination of learning, planning, and acting.

As famous author Alan Lakein once said, "Planning is bringing the future into the present so that you can do something about it now."

Dr. Wesley E. Donahue
2023

MASTERNING PROJECT MANAGEMENT

Planning for Performance

What Readers are Saying

What you don't know can hurt you and your organization if you don't plan. Doing the work might be 90 percent of the job, but failing to plan can tank even a brilliant project. After completing this interactive series, I was able to complete multiple big projects in less time and with less stress.

I didn't know what a work breakdown structure was before reviewing these materials. Now I don't know how any project manager can develop a scope statement without preparing one.

The examples opened my eyes to the importance and benefits of formalizing many of our project management forms, templates, and processes.

Many project management books focus on complicated tools and formulas that aren't really practical to use in many work environments. This book is different and stresses that "simple is good!"

Many of the tools covered in this book complement what our organization is doing with "Lean" and our operational excellence initiatives.

With heaps of know-how, experience, and research data to back it up, Dr. Donahue has produced a fantastic resource for anyone who wants to be a better project manager. We can directly apply the tools we learned on our projects.

Acknowledgments

We did not create this book alone. I want to acknowledge the people who contributed significantly to the work: Drs. Lisa Donahue and Rebecca Sarnaski for their invaluable research efforts in tracking down articles, books, and other sources needed for content development and identifying examples; Dr. Lisa Turner and Billie Tomlinson for their review and editing of content presented in this book; Valentine Platon and Sommoade for adding graphics that help bring the text to life; Zizi Iryaspraha Subiyarta and Alex Donahue for thoughtful design suggestions; Amit Dey for text formatting; former instructors and professional associates at Penn State Management who shared their business and industry wisdom and years of teaching experience; the individuals who took the time to review the manuscript, and the many thousands of people who participated in surveys, focus groups, and interviews, without whom this workbook would lack the richness of real-world detail. To all these people I offer a sincere and heartfelt thank you!

Contents

A clear vision, backed by definite plans,
gives you a tremendous feeling of confidence and personal power.

—Brian Tracy

Introduction

Audience

Mastering Project Management is a three-part book series designed for individuals who manage or anticipate managing workplace projects such as technical specialists, apprentices, frontline leaders, engineers, managers, and project stakeholders.

Book 1, ***Planning for Performance,*** focuses on the technical side of the traditional project management life cycle and covers the tools and techniques most widely used to manage workplace projects. This book covers the competency areas of Planning and Evaluation, Resource Management, Problem Solving, Communication, Leadership and Coaching, and Management Controls and are covered in the four chapters depicted in the table below.

The two companion books, ***Results Through People,*** and ***Controlling Time, Money, and Risk,*** are focused on the leadership and business sides of project management, respectively.

Mastering Project Management		
Book 1 **Planning for** **Performance**	**Book 2** **Results Through People**	**Book 3** **Controlling Time,** **Money, and Risk**
• Contemporary Project Management • Defining Project Scope • Project Planning and Scheduling Tools • Monitoring and Controlling Projects	• Enhancing Project Communication • Developing Interpersonal Project Relationships • Building High-Performance Project Teams • Leading and Managing Projects	• Identifying and Mitigating Project Risk • Controlling Project Costs and Budgets • Managing Time and Project Meetings • Completing and Closing Projects

All three books are aligned with the Project Management Institute's (PMI)® Project Management Body of Knowledge (PMBOK® Guide) and are competency-based focusing on sets of skills, knowledge, attitudes, and behaviors that are observable and measurable.

As a bonus, each of the 12 chapters in the series contain a 20-question knowledge review test with answers for individuals interested in potentially pursuing certification through the Project Management Institute or obtaining Google's Project Management Certificate.

Structured Learning Design

All our books and workbooks align with the research-based **Competency Model.** The model, which is rooted in work by the U.S. Department of Labor and others, gives you a framework for structured learning by helping you identify and develop specific competencies.

If you ask people to define competency, you may be amazed at the variety of responses you receive. We define a *competency* as a set of skills, knowledge, attitudes, and behaviors that are observable and measurable. *The emphasis here is on observable and measurable.* It is not enough that you *think* you are competent in an area. You must be able to *behave* in ways that demonstrate your competence to others.

Our framework has **35 competency dimensions** associated with successful performance in leadership and professional roles.

Centrestar Competency Model

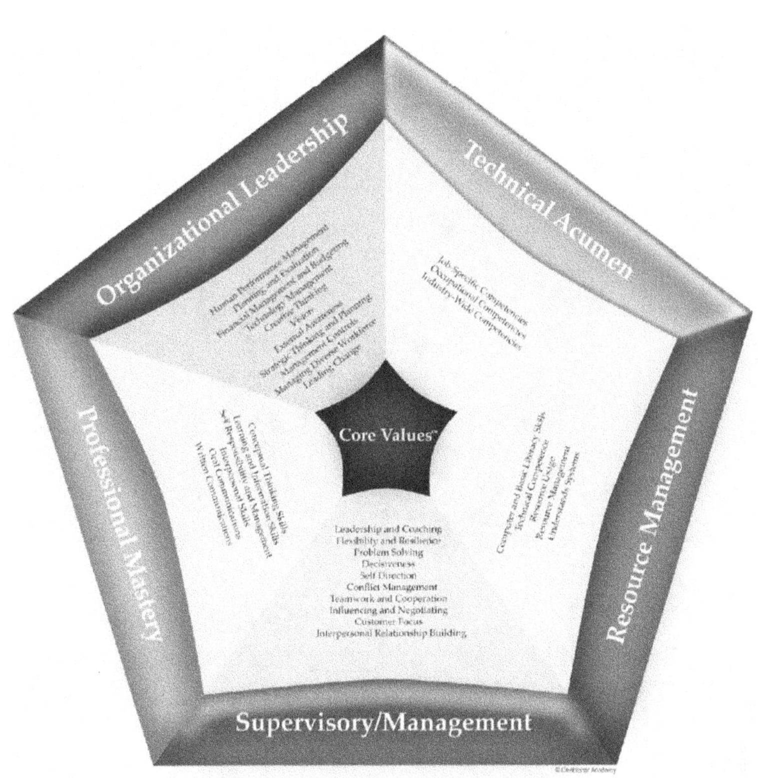

Based on thousands of business participant responses from many industries, we clustered these 35 competencies into five competency domains, which we named and coded as follows:

- A.   Resource Management
- B.   Professional Competence
- C.   Supervisory/Management
- D.   Organizational Leadership
- E.   Technical Acumen

Competencies and clusters may overlap. The content of this **Planning For Performance** book are most closely associated with the competency clusters indicated below.

A. RESOURCE MANAGEMENT
_____ 1. Computer and Literacy Skills
_____ 2. Technical Competence
_____ 3. Resource Usage
_X___ 4. Resource Management
_____ 5. Understands Systems

B. PROFESSIONAL COMPETENCE
_____ 6. Conceptual Thinking
_____ 7. Learning and Information Skills
_____ 8. Self-Responsibility and Management
_____ 9. Interpersonal Skills
_X___ 10. Oral Communication
_X___ 11. Written Communication

D. ORGANIZATIONAL LEADERSHIP
_____ 22. Human Performance Management
_X__ 23. Planning and Evaluation
_____ 24. Financial Management and Budgeting
_____ 25. Technology Management
_____ 26. Creative Thinking
_____ 27. Vision
_____ 28. External Awareness
_____ 29. Strategic Thinking and Planning
_X__ 30. Management Controls
_____ 31. Diverse Workforce
_____ 32. Leading Change

Centrestar Competency Model

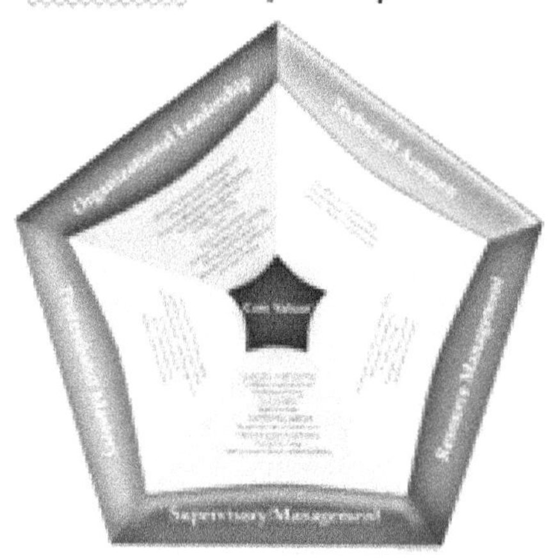

C. SUPERVISORY MANAGEMENT
_X__ 12. Leadership and Coaching
_____ 13. Flexibility and Resilience
_X__ 14. Problem Solving
_____ 15. Decisiveness
_____ 16. Self-Direction
_____ 17. Conflict Management
_____ 18. Teamwork and Cooperation
_____ 19. Influencing and Negotiating
_____ 20. Customer Focus
_____ 21. Interpersonal Relationship Building

E. TECHNICAL ACUMEN
_____ 33. Job-Specific Competencies
_____ 34. Occupational Competencies
_____ 35. Industry-Wide Competencies

This ***Planning For Performance*** *guide*book focuses on the competency areas of:

- **Resource Management.** Demonstrates awareness of technical resources; knows how to apply resources to achieve desired outcomes.

- **Oral Communication.** Makes clear and effective presentations to individuals and groups; listens to others.

- **Written Communication.** Communicates effectively in writing; can critically review and comprehend information written by others.

- **Leadership and Coaching.** Models and encourages high standards of ethical behavior; adapts leadership styles to situations and people; empowers, motivates, guides, and coaches.

- **Problem Solving.** Recognizes and defines problems; analyzes relevant information; encourages alternative solutions, develops plans to solve problems.

- **Planning and Evaluation.** Establishes policies, guidelines, plans, and priorities; plans and coordinates with others; aligns required resources; monitors progress and evaluates outcomes; improves organizational efficiency and effectiveness.

- **Management Controls:** Ensures the integrity of the organization's processes; promotes ethical and effective practices.

How to Use This Book

This book is structured to be a hands-on guide with options for how you can use the material.

One way to use the book is to read it straight through. Or you can jump to specific chapters depending on your interests and goals. We recommend you start by reviewing the table of contents, so you understand the workbook's organization. Read the overview of each chapter and then complete the introductory assessment by rating your level of agreement with the ten statements. This will help you clarify your current thinking. After that, visually scan the book. Scanning will help you pinpoint areas where you may have an information gap and areas where you feel confident. From there, you can set your learning goals and dive deeper into the material.

The book will help you develop your competence by addressing the ten most vital concepts associated with each of the four topical chapters. The content for the ten concepts is structured in a consistent learning format.

Each concept starts with several paragraphs of relevant content and ends with three **What to Do** action suggestions, followed by a **Remember** section which lists important learning points. To **Enhance Your Learning**, we offer links to additional resources. Each concept area concludes with a **Reinforce Your Learning** application activity.

At the end of each topical chapter is a **Summary** and a **Recap Checklist**. The recap lists all 30 of the **What to Do** actions from each concept in the workbook.

As with most things in life, you will get out of this book only what you put into it. To learn and grow, you must engage fully with the material presented. Ask yourself questions as you read the material in the book. Do the activities and answer the questions in each concept. Make notes. Look for ideas new to you and consider how they fit with your current knowledge. Recognize what you already know and can build on, and what might be a new way of looking at something.

This book aims to ensure that your skills are at the highest level they can be. Engaging with the concepts and materials will help you become a more rounded professional who experiences success in your chosen career.

Chapter 1

Contemporary Project Management

Plans are nothing; planning is everything.

—Dwight D. Eisenhower

Overview

Contemporary Project Management

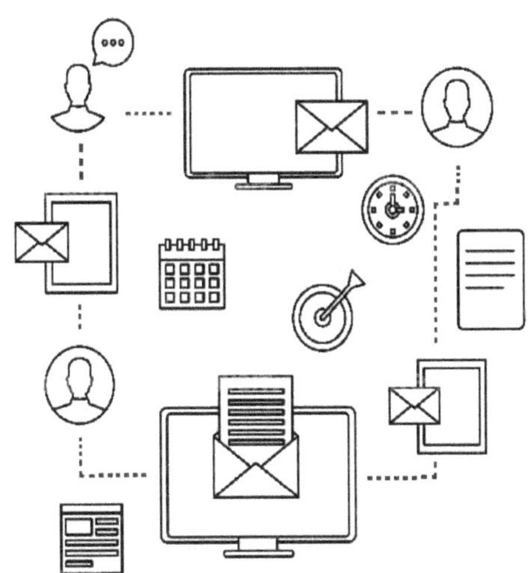

"Plans are nothing; planning is everything."
—Dwight D. Eisenhower

A *project* is a temporary work assignment to achieve a specific goal or outcome. It is essential to recognize that the project life cycle includes the phases of beginning, planning, doing the work, and closing, and that people's actions may account for the majority of project problems. These problems include issues related to the management of the project.

This chapter describes the project management process and examines project problems, issues, and challenges with a focus on the human side. Understanding the roles, responsibilities, and attributes of project managers is essential for selecting individuals with the appropriate qualifications.

In this chapter, you will learn how to:

- Distinguish project management from conventional management
- Describe the general phases of a project
- Understand the importance of a formalized project management methodology
- Recognize the attributes of a project manager
- Understand the key roles of project managers and the responsibilities associated with each

The competencies associated with this chapter include:

Planning and Evaluation   **Resource Management**   **Leadership and Coaching**

Take Your Temperature for Contemporary Project Management

With yourself and your organization in mind, read each statement carefully. Next to each statement, write the number (from 1 to 10) that indicates your level of agreement with the statement.

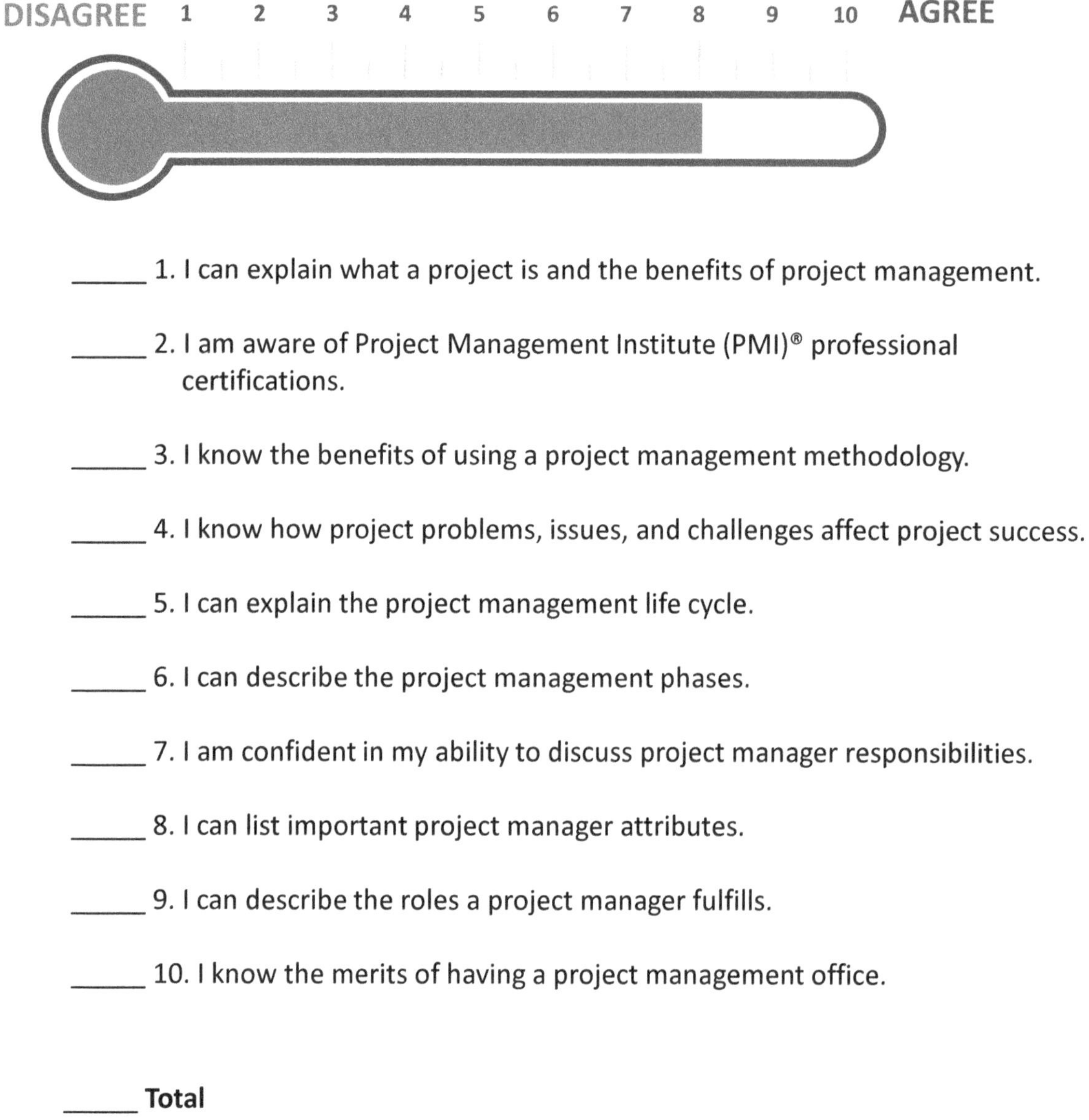

______ 1. I can explain what a project is and the benefits of project management.

______ 2. I am aware of Project Management Institute (PMI)® professional certifications.

______ 3. I know the benefits of using a project management methodology.

______ 4. I know how project problems, issues, and challenges affect project success.

______ 5. I can explain the project management life cycle.

______ 6. I can describe the project management phases.

______ 7. I am confident in my ability to discuss project manager responsibilities.

______ 8. I can list important project manager attributes.

______ 9. I can describe the roles a project manager fulfills.

______ 10. I know the merits of having a project management office.

______ **Total**

Take a moment to reflect on the following before you move forward in the chapter.
In your work, what problems have you experienced or observed related to project management?

Take a few minutes to reflect on your self-assessment. List two or three areas you want to improve.

As you progress through this chapter consider what actions you can take to demonstrate competence in the following three competency areas:

1. **Planning and Evaluation**. Establishes policies, guidelines, plans, and priorities; plans and coordinates with others; aligns required resources; monitors progress and evaluates outcomes; improves organizational efficiency and effectiveness.

2. **Resource Management.** Demonstrates awareness of technical resources; knows how to apply resources to achieve desired outcomes.

3. **Leadership and Coaching.** Models and encourages high standards of ethical behavior; adapts leadership styles to situations and people; empowers, motivates, guides, and coaches.

Define Project and Project Management

According to the **Project Management Institute (PMI)®** and **A Guide to the Project Management Body of Knowledge (PMBOK)®**,[1] *project management* is the application of knowledge, skills, tools, and techniques to project activities to meet project requirements. A *project* is a temporary endeavor undertaken to create a specific product, service, or result.

Project management differs from conventional management in that projects have a finite life with a specific beginning and end.

Examples of Activities Associated with Projects and Project Management	Examples of Activities Associated with Conventional Management
Constructing a building	Maintaining a building
Relocating an office	Managing an office
Installing a piece of equipment	Operating equipment
Developing a strategic plan	Leading an organization
Developing a training program	Managing performance

Here are the typical characteristics of most projects:
- Limited time span
- Led by a project manager
- Completed by a team whose job is in their functional area, so team members report back to their operational areas after the project is over, and they sometimes work on a project part-time
- Constrained by human performance, time, and budget
- Organized to achieve a specific goal (product, service, or result)

Program management is the management of a group of related projects. It may involve sharing resources between projects. The work product of one project may affect associated projects.

[1] PMI and *PMBOK* are registered marks of the Project Management Institute, Inc.

Portfolio management typically includes the management of multiple programs. It is the most complex level in the project management hierarchy because of the number of programs and associated projects.

The graphic below illustrates the **Traditional Project Management Hierarchy**, with the broadest or most encompassing level at the top and moving to the most specific level, which is the management of a single project. Notice the "we are here" arrow. It highlights that this workbook is primarily concerned with the management of one project and *not* with program or portfolio management.

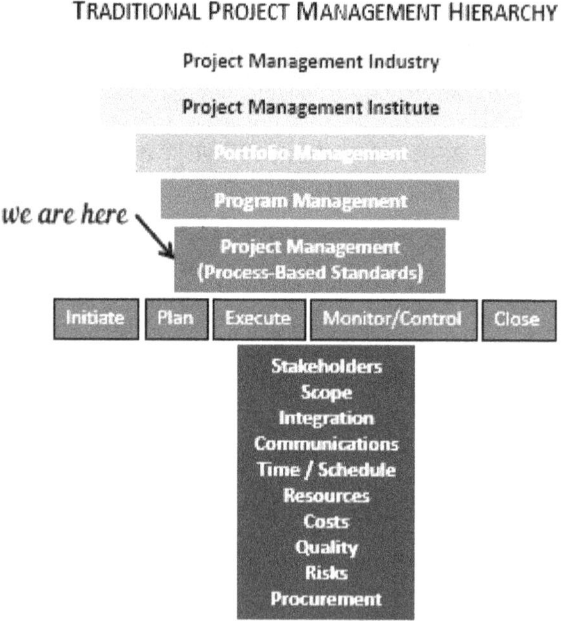

Also notice that project management is divided into major phases: Initiate; Plan; Execute; Monitor & Control; Close. Note that managing a project requires consideration of ten related knowledge or process areas. These knowledge and process areas are stakeholders, project scope, integration, communications, time and schedule, resources, costs, quality, risks, and procurement. These ten areas are the ten *knowledge areas* within the Project Management Body of Knowledge (PMBOK® Guide, 6th Edition). We will discuss these knowledge areas throughout this course.

Recent Developments by the Project Management Institute (PMI)®

The Project Management Institute (PMI) has also recently developed standards (PMBOK® Guide, 7th Edition) by which project management may be viewed from a couple of additional perspectives. In addition to looking at **project management** from a **process-based** standards orientation, project management may be viewed from a **principles-based** orientation as well as a **systems-based** orientation. For the process-based orientation, the focus is on the deliverables; for the principles-based orientation the focus is on outcomes, and for the systems-based orientation, the focus is on linking business capabilities to meet an organization's strategy, value, and business objectives.

These additional orientations are not meant to negate the process-based standards orientation, but rather to compliment them. The additional orientations add a broader global business perspective to the field of project management by permitting a shift in thinking from the traditional ten knowledge areas (illustrated in the previous graphic) towards the additional twelve guiding project management principles and eight project performance domains as depicted in the **Contemporary Project Management Hierarchy** graphic that follows. The additional orientations may be particularly useful from the perspective of governing larger projects, programs, and portfolios, and linking them to overarching business capabilities, strategies, and objectives.

Contemporary Project Management Environment

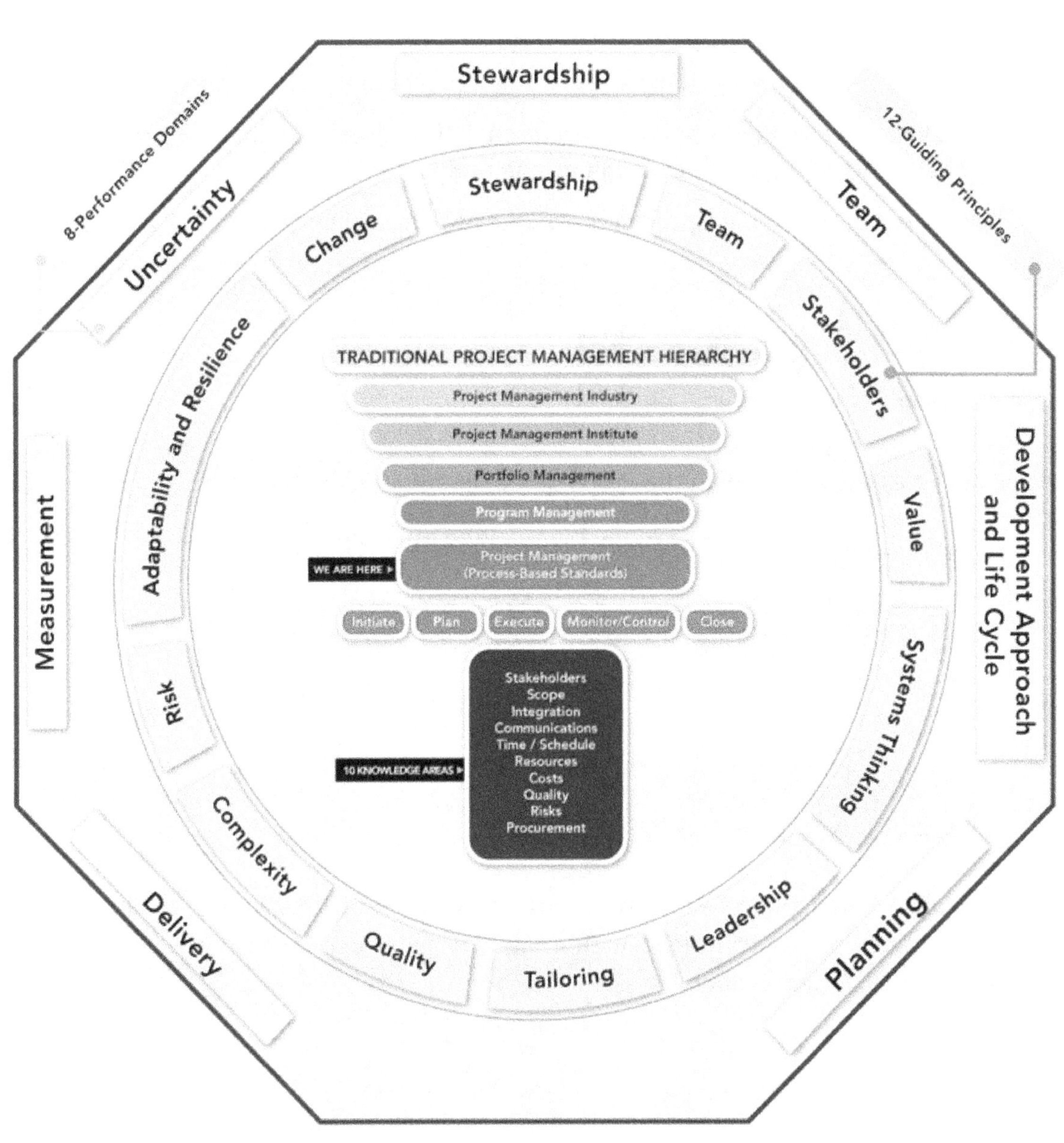

What to do:

☐ Review the Project Management Institute (PMI)® website (www.pmi.org).

☐ Identify some of the projects you have been involved in or observed.

☐ Brainstorm issues, problems, and training needs related to those projects.

Other actions:

☐ _______________________

☐ _______________________

☐ _______________________

☐ _______________________

Remember

✓ A project is a temporary endeavor undertaken to create a product, service, or result.

✓ Project management differs from conventional management in that projects have a finite life cycle with a specific beginning and end.

✓ The process of managing a project is divided into phases: Initiate, Plan, Do, Close.

Enhance Your Learning

Review activities associated with the Project Management Institute (PMI)® at:

Introduction to Project Management Institute (PMI)®

PMI Certifications

PMI - PMBOK® Guide & Standards

Reinforce Your Learning

Typical Project Management Problems. Consider typical problems, issues, challenges, and needs related to project management that you have experienced or observed. Take a few minutes to list them below.

Become Familiar with the Project Management Institute (PMI)®

Project Management Institute (PMI)®

The project management industry is a global collection of project management professionals and organizations and the activities associated with their work. The **Project Management Institute (PMI)®**, founded in 1969 and with over 500,000 members worldwide, is a premier organization for professional project managers. PMI sets standards and delivers seminars and educational programs, and it offers professional certifications.

A Guide to the Project Management Body of Knowledge (PMBOK)®

The PMI® organization works with volunteers to create industry standards, such as *A Guide to the Project Management Body of Knowledge (PMBOK)®*. This guide is in its 7th edition, released in the fall of 2021. However, most of the content in the 6th edition is still relevant and referred to in the 7th edition.

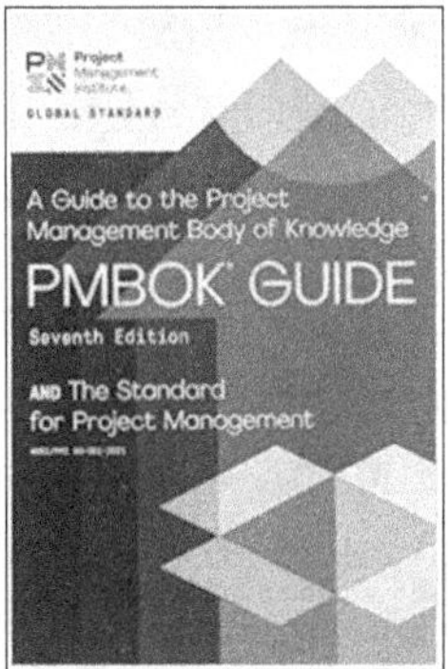

Past editions of the *PMBOK® Guide* (including the 6th edition) have represented a process-based standard for project management. In these guides, inputs, tools, and techniques, and outputs related to the project management processes are discussed, and where the outputs of one process can be the inputs to another process.

Due to the continually evolving landscape of project management, the Project Management Institute (PMI)® introduced the 7th edition which incorporated, in addition to the traditional process-based standards, an orientation towards a principles-based standard with a focus on outcomes instead of deliverables and incorporated a broader systems-view of project management.

PMI Certifications

PMI offers specialized certification in several areas of project management to support individuals in developing skills in related areas. Like a degree or other focused training, a PMI certification shows that the holder has the skills necessary to do a professional job. The following are some of the most common certifications PMI offers:

- Portfolio Project Manager (PfMP)®
- Program Management Professional (PgMP)®
- Project Management Professional (PMP)®
- Certified Associate Project Manager (CAPM)®
- Business Analyst (PMI-PBA)®
- Agile Certified Practitioner (PMI-ACP)®
- Risk Management Professional (PMI-RMP)®
- Scheduling Professional (PMI-SP)®

The **Portfolio Management Professional (PfMP)®** certification is for professionals who manage groups of projects as they relate to an organization's strategic goals. A PfMP® can manage one or more portfolios.

Requirements for the PfMP® include a panel review; passing a 4-hour, 170-question multiple-choice exam; and **either** a bachelor's degree (or global equivalent) with a minimum four years (6,000 hours) of portfolio management experience and eight years (96 months) of professional business experience within the last 15 years, **or** a high school diploma (or global equivalent) with a minimum of seven years (10,500 hours) of portfolio management experience, and eight years (96 months) of professional business experience, within the last 15 years.

The **Program Management Professional (PgMP)®** certification shows the holder is a senior-level practitioner who has the skills to manage related projects and to coordinate them where possible. Program-level management facilitates a synergism among projects that would otherwise not be possible.

Requirements for the PgMP® certification include a bachelor's degree (or global equivalent) with a minimum of four years (6,000 hours) of project management experience **and** seven years (10,500 hours) program management experience within the last 15 years; a panel review; and passing a 4-hour, 170-question, multiple-choice exam.

The **Project Management Professional (PMP)®** certification shows the holder has the skills and knowledge to manage projects from beginning to end. It requires passing an exam that tests a candidate's working knowledge of professional practices. The certification usually helps the holder earn more money.

Requirements for the PMP® certification include 35 hours of professional project management education; passing a 4-hour, 200-question, multiple-choice exam; and **either** work experience to include a bachelor's degree (or global equivalent) and a minimum of three years (4,500 hours) leading projects, **or** a high school diploma (or global equivalent) and a minimum of five years (7,500 hours) leading projects.

The **Certified Associate Project Manager (CAPM)®** is for people who are involved or associated with projects but not currently working as a project manager. Examples are a project team member, a project stakeholder, or a subject matter expert (SME) who works with project teams. This certification shows the holder is aware of project management terminology and methodologies.

Requirements for the CAPM® includes a high school diploma, associate degree (or global equivalent); passing a 3-hour, 150-question, multiple-choice exam; and **either** at least 23 hours of project management education without professional work experience; **or** at least 1,500 hours of experience on a project team without project management education.

The **PMI Professional in Business Analysis (PMI-PBA)®** certification addresses the need for professionals who have the skills to identify and document project requirements. Project requirements are the foundation of the project scope, which is used to develop the project plan. The ability to capture requirements accurately and efficiently is a significant benefit to a project team. Areas of focus include outputs and projected business results.

Requirements for the PMI-PBA® include 35 hours of business analysis education, passing a 4-hour, 200-question multiple-choice exam; and **either** a bachelor's degree (or global equivalent) with minimum three years (4,500 hours) of business analysis experience and 2,000 hours working on project teams within the last eight years; **or** a high school diploma (or global equivalent) with a minimum five years (7,500 hours) of business analysis experience and 2,000 hours working on project teams within the last eight years.

The **PMI-Agile Certified Practitioner (PMI-ACP)®** certification shows the holder has knowledge of the principles and techniques associated with the Agile project management method. The education covers various approaches to Agile, including Lean, Scrum, Kanban, extreme programming (XP), Crystal, and test-driven development (TDD).

Requirements for the PMI-ACP® include a minimum of 21 hours of professional education in Agile practices; 2,000 hours working on project teams in the last five years; **or** 1,500 hours working on Agile project teams or with Agile methods in the previous three years.

The **PMI Risk Management Professional (PMI-RMP)®** certification focuses on the skills needed to identify and mitigate negative risks while leveraging positive risks. As project environments become increasingly complex, risk management expertise has become more necessary to project success, and it is a specialized skill. Organizations that practice professional risk management are more successful at completing projects within their constraints.

Requirements for the PMI-RMP® certification include **either** a bachelor's degree (or global equivalent) with a minimum of 30 hours in project risk management; **or** a bachelor's degree (or global equivalent) with a minimum 3,000 hours of work experience in project risk management within the last five years.

The **PMI Scheduling Professional (PMI-SP)®** certification shows the holder has the skills to manage all aspects of project scheduling. Task duration estimates, task relationships, interactions with shared resources, and global concerns are all considerations in developing and maintaining a schedule.

Requirements for the PMI-SP® certification include passing a 3.5 hour, 170-question multiple-choice exam; **either** a bachelor's degree (or global equivalent) with a minimum 30 hours in project scheduling **or** a high school diploma (or global equivalent) with a minimum 40 hours in project scheduling; and **either** a bachelor's degree (or global equivalent) with minimum 3,500 hours of work experience in project scheduling within the last five years **or** a high school diploma (or global equivalent) with a minimum 5,000 hours of work experience in project scheduling within the previous five years.

PMP® Examination

The PMP® examination is the most popular by number of people certified and validates a candidate's knowledge in the domains of people, process, and business environment. According to PMI, the following is an approximate breakdown of the percentage of questions in each of the three groupings:

Domain	Percentage of Items on Exam
People	42%
Process	50%
Business Environment	8%
TOTAL	100%

While various industries may use different terminology, the test focuses on the universal knowledge areas necessary to manage projects successfully in any organization. For example, consulting engineers may use the term "subcontracting" as opposed to "procurement planning." But across industries, good management practices are the same and the differences are semantic.

After Certification

After attaining the Project Management Professional Certification, a PMP® must maintain status by satisfying the requirements of the Professional Development Program. Certified project managers must complete no less than 60 Professional Development Units (PDUs) within a three-year cycle and agree to continue to adhere to the PMI Code of Professional Conduct.

You can find more information about the Professional Development Program and Professional Development Units on the PMI website.

What to do:	**Other actions:**
☐ Discuss project management certifications with your manager for applicability.	☐ ______________________ ☐ ______________________
☐ Find a certified associate and ask them about the advantages of being certified.	☐ ______________________ ☐ ______________________
☐ Pursue one of the certifications, if appropriate.	☐ ______________________

Remember

✓ PMI offers many educational opportunities and several types of professional certifications. Some of these may be of benefit to you in your job.

✓ Explore the PMI website and talk to people in your organization who are familiar with PMI.

✓ Consider some projects in your organization and the kinds of skills needed to manage those projects.

Enhance Your Learning

The following link provides more information about the Project Management Professional (PMP)®:

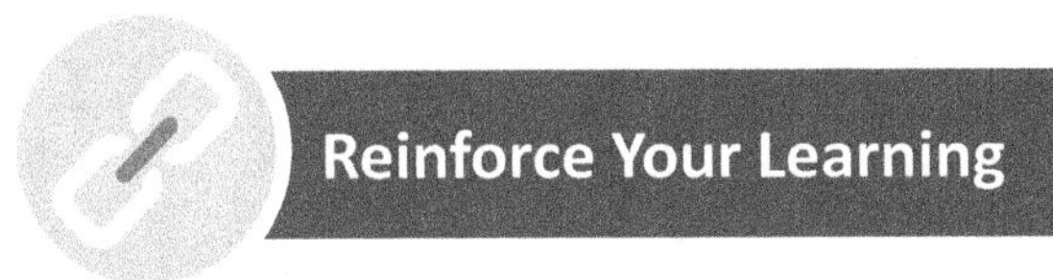

https://www.pmi.org/certifications/project-management-pmp

Reinforce Your Learning

Benefits of Project Management Certifications. Consider projects you have been associated with or observed. Describe how the skills gained in PMI certifications may have helped prevent problems or challenges on these projects?

22

Why do so many professionals say they are project managing, when what they are actually doing is firefighting?

—Colin Bentley

Contemporary Project Management

Know the Benefits of Using a Project Management Methodology

The payoff of following a project management methodology is great. Organizations benefit from the efficient and effective creation of a new product, service, or result. Stakeholders benefit from achieving their business and operational goals. Project team members benefit through the satisfaction of a job well done and by gaining experience that will help them on the next project and in building their careers.

On the flip side, those who do not use a project management methodology are more likely to fail in delivering a product, service, or result. According to *Entrepreneur Magazine*, here are **five mistakes** that **cause projects to fail** (2015):

1. Eyes bigger than your budget.
2. Working with the wrong talent.
3. Continuing to pursue bad ideas.
4. Death by committee.
5. Keeping the final approver in the dark.

You can avoid or mitigate these problems by adhering to an established project management methodology. However, there are a number of popular project management methodologies employed in business and industry today. Most follow a system of practices, techniques, procedures, and rules used by those who work in various disciplines. Some of the popular project management methodologies include:

- **Agile** fosters iterative development and team collaboration.

- **Scrum** is a flexible iterative cross-functional "sprint event" approach.

- **Kanban** focuses on visibility of work-in-progress.

- **Lean** focuses on streamlining and getting rid of waste.

- **Waterfall** is a linear sequential approach where progress flows downward.

- **Six Sigma** aims to improve quality by reducing errors.

- **Prince2** defines inputs and outputs for every stage of a project.

- **PMI/PMBOK®** involves using five process groups (initiating, planning, executing, monitoring and controlling, and closing).

The Methodology Does Not Ensure Success

No matter the methodology, size, or complexity, how forgiving or tight the schedule, the skills of the team members, or the commitment of the stakeholders, all projects have one thing in common. This essential factor is that without careful planning, a project will be plagued by one or more of the following problems:

- Delays and missed deadlines
- Wasted time and resources
- Cost overruns
- Inadequate quality
- Conflicts

Poor project planning can sink a project. However, you can avoid or mitigate these problems by adhering to an established project management methodology. Here are a few examples of the benefits of using a project management methodology:

Project goals will be clearer.

Duplication of effort which could result in cost overruns will be lessened.

Potential issues can be identified before causing unnecessary problems.

Potential negative events can be anticipated, and the risk planned for.

Schedules and deadlines can be maintained.

Potential litigation can be avoided from negative events or issues.

Morale can be maintained by building teams with the right people.

Workloads can be balanced.

Almost any major project will cost more, take more time, and have more problems than it would have if effectively planned. This means following a proven project management methodology. Planning is at the heart of all successful projects.

What to do:	**Other actions:**
☐ Consider the benefits of following a step-by-step project management methodology.	☐ ______________________________
	☐ ______________________________
☐ Identify the sources/kinds of information you need access to.	☐ ______________________________
	☐ ______________________________
☐ Reflect on the risks in your project and what their impact could be.	☐ ______________________________

Remember

✓ Following a project management methodology improves communication on a project. Everyone has the same expectations and shares the same information.

✓ We have all seen what happens when projects go wrong. Problems are less likely to occur when everyone follows the same project management process, and when problems arise, they are easier to understood and correct.

✓ Communication is a critical part of any project.

Enhance Your Learning

Watch the following 3-minute video that overviews the importance of project planning by Andy Kaufman. Leading cause of project failure: Poor project planning.

Kaufman, A. (2010). *Leading cause of project failure: Poor project planning.*		Available at: http://www.youtube.com/watch?v=MwQ_5JfDIJs

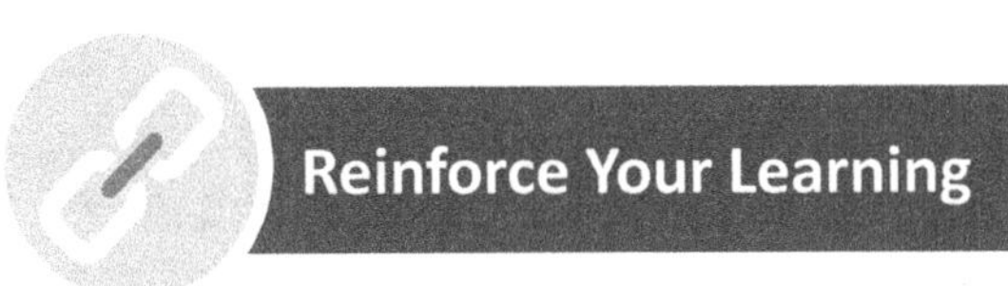

Reinforce Your Learning

Identify Causes of Project Failure. In your experience, what are the most common causes of project failure?

Be Aware of How Projects Begin, the People Involved, and Quality

Organizations move forward based on the success of the projects they choose to undertake and how successfully they complete them. Projects start as ideas championed by various people in an organization, and these people typically expect high-quality results when ideas become projects.

In this concept, we will discuss how ideas become projects, the people involved in projects, and how project managers are responsible for delivering quality.

Project Ideas

Ideas for projects come from many sources, and projects can be internal to an organization or external. For example, you work in your organization's information technology department and your department provides project services to develop and install business software for the organization. Or your organization might contract a project with another organization to build commercial office space.

Anyone can suggest an idea for a project. For example, an idea might come from a senior manager who wants to support a new business objective, an operations manager who needs more space or to improve a business process, or from a technical person who has an idea for a better way of doing something.

Organizations gather project ideas and then evaluate them to identify the projects that will bring the greatest benefits. Most organizations do not have the resources to develop all projects at once, so they plan and prioritize projects to achieve the maximum benefits to the business.

How Organizations Review Project Ideas

Organizations review ideas for projects in many ways, but most include an assessment of the benefits the project could bring to the organization. Some organizations do formal studies. For example, they may do a feasibility or return-on-investment (ROI) study. The purpose is to better describe the project, and so to better calculate the resources, costs, and time needed to complete the project. Or, some organizations may do studies during the first phase of the project, which is the planning phase, to clearly define the project's requirements and costs

Once an idea for a project is approved, larger organizations may go a step further and formally compare one project to another to determine which project has priority. For example, an organization may decide that Project A will result in added revenues immediately, while Project B will not produce new revenue until the following year. The organization may then choose to do Project A first, and once it is underway or complete, they will do Project B.

People Involved in a Project

Each project has a sponsor who formally presents an idea for a project and approves the initial budget for the project. Sponsors are typically senior managers, and they may not have much day-to-day involvement in the project after it begins.

Approved projects have someone named as the primary client. The client is one of the stakeholders but is responsible for the business oversight of the project as it progresses. For example, the client may be an operational manager whose business area will have the most to gain when the project is complete. Or the client could be a manager who works for an outside organization the project has contracted with to provide a product, service, or result. The client has the final say in any necessary business decisions as the project progresses. For example, the client must give the final approval for any changes to the project once it is underway because changes usually affect project costs.

People who helped champion the project idea may become project stakeholders. According to PMI, a stakeholder is anyone affected by a project. For example, besides the project's client, stakeholders could include the sponsor, other executives, senior managers, operational managers, end-users, and even members of the public.

Like the client, stakeholders have an interest in the project's product, service, or result, and they must be kept informed of status and issues as the project progresses. Stakeholders often assist the client in making project business decisions, for example, analyzing proposed changes, helping solve problems with vendors, or determining a business solution for a project risk.

The project manager is responsible for project integration and putting all the pieces of a project into a cohesive whole. This includes planning, managing (monitoring and controlling), and communicating all day-to-day project activities throughout the complete life cycle of the project. The project manager is a professional who serves as the focal point of the project, the person who needs to have a handle on everything happening, and who communicates project information with the client and other stakeholders, the team, and any external people involved in the project.

The project team is responsible for doing the work of the project. The team may or may not formally report to the project manager, though team members are accountable for the tasks assigned by the project manager.

Sometimes the project manager picks team members, but sometimes team members are assigned to the project by the organization. They may work on the project full-time or part-time, depending on the project's needs for the skills they provide.

Quality

Quality is a goal in all organizations. One definition of *quality* is that it measures the degree to which a process, product, service, or result conforms to established requirements.

Quality in a project is often assessed by the stakeholders based on their expectations about the product, service, or result the project will provide. For example, an operations manager in a warehouse may want a new software system to better support the warehouse's business processes.

Stakeholder expectations are not always clear to the project manager and team at the start of the project. But the project may not be judged to be successful if it does not satisfy the expectations.

To deliver a successful project, the project manager must work to understand all stakeholder expectations, ensure the project meets those expectations or if not, that the stakeholders are aware of the reasons why not.

The project manager is responsible for defining quality standards during the planning phase, and for monitoring and controlling quality over the life of the project.

Projects are judged to be successful in terms of quality when the following criteria are satisfied:

- Business objectives are met
- Scope requirements are met
- Completed within budget
- Schedule is met
- Client is satisfied

What to do:

☐ Identify the criteria your organization uses to judge project success.

☐ Define what quality means for projects you (or your organization) have completed.

☐ Complete an Idea Appraisal Worksheet for a project idea that you or someone in your organization is considering.

Other actions:

☐ _______________________

☐ _______________________

☐ _______________________

☐ _______________________

☐ _______________________

✓ Almost any project will cost more, take more time, and create more problems than it would have if effective project management had been used.

✓ At the beginning of a project, reviewing historical records and seeking advice from those who managed similar projects in the past is a good idea.

✓ The potential payoff of following a step-by-step planning process is high.

✓ Evaluate the merits of various project ideas before investing effort and resources. Consider ranking ideas and potential projects to identify those with the greatest benefit to the organization.

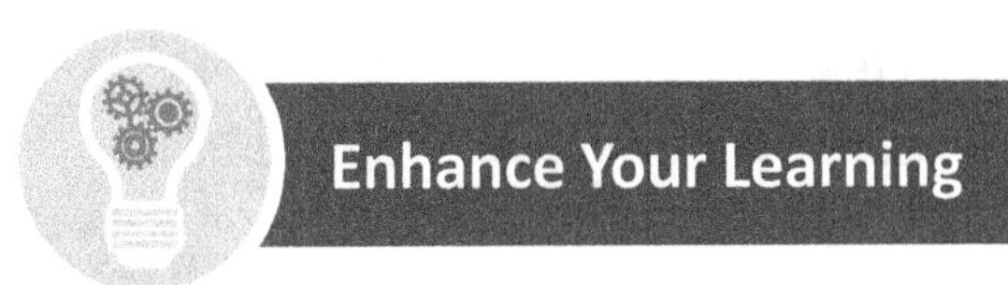

Review the article by ProjectManagement.com to learn more about the reasons why projects fail.

Project Management.com. (2021). *Top 10 Reasons Why Projects Fail.*		Available at: https://project-management.com/top-10-reasons-why-projects-fail/

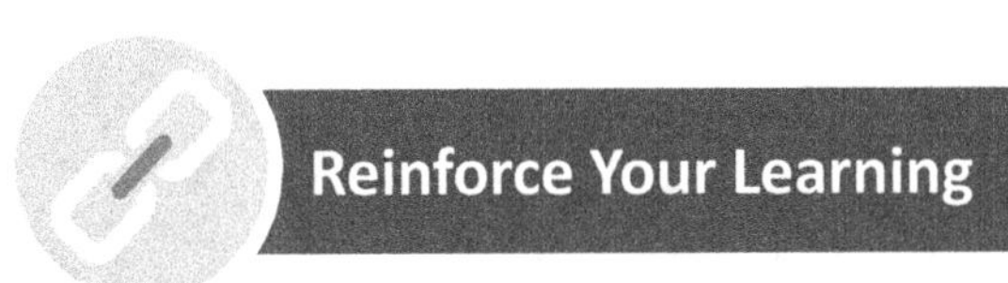

Project Idea Appraisal Worksheet. Projects start with ideas. A manager may wonder if relocating a warehouse would save money; a product manager may want to attract new business by adding features to a software product, or you may need to plan a business trip.

Identify a project that you or someone in your organization is considering. With that project in mind, complete the **Idea Appraisal Worksheet** below. You will find a copy of the worksheet in **Appendix A** that you can photocopy for use with other ideas.

For more sophisticated project ideas, you may need build a business case to warrant further examination. If so, there are a number of free Project Management templates available. Here is a link to a free **Business Case Template** available from Project Management Docs:

https://psu.instructure.com/courses/2185007/pages/project-management-docs-templates-%7C-focus-on-application?module_item_id=35373675

Project Idea Appraisal Worksheet

Use this worksheet to vet or evaluate an idea for a project by focusing on the benefits the completed project would give the stakeholders and the organization.

In a few words, what is the project idea?

What do you hope to accomplish with a project based on this idea?

What is the project's tangible deliverable? This is a product, service, or result.

What is your rough estimate for how much the project might cost?

How long might the project take?

What kind of resources—people skills and material resources—would you need?

If needed, who might be willing to back the project financially?

Who is the client (the primary stakeholder) and who are the other stakeholders?

So, do you want to pursue this project? Why, or why not?

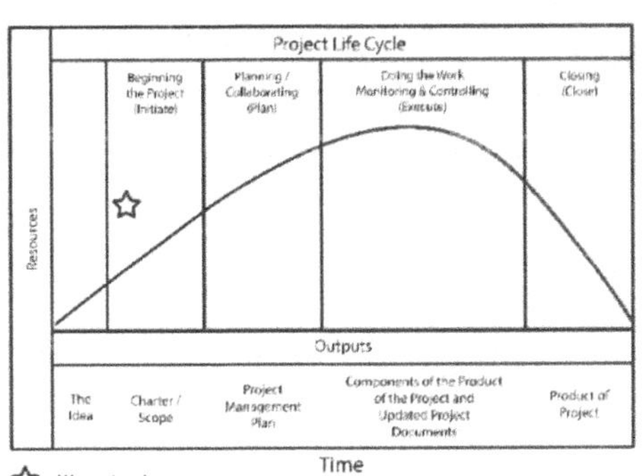

Explore the Project Life Cycle

The Project Life Cycle illustration below shows the phases, the graduated use of resources, and the outputs in the life cycle of a typical project.

The phases in the project life cycle include beginning, planning, doing the work while monitoring and controlling, and closing.

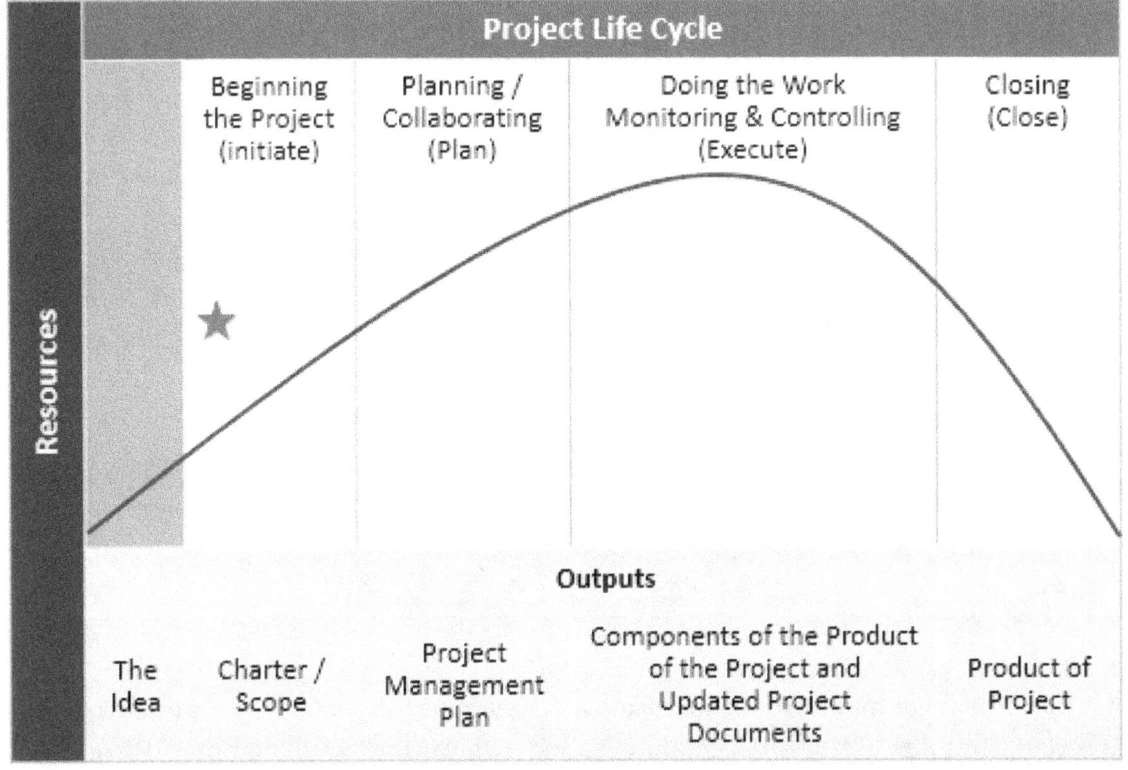

The curved line indicates the number of resources typically expended in each phase. It shows that the beginning of a project uses limited resources to establish the requirements and that more resources are needed as the project progresses. The execution phase requires the most resources. As the project approaches the end, resource needs decrease because many activities are complete. The exact shape of the curve can vary, depending on the type of project.

Each phase has processes associated with it and essential project documents.

A **project manager** is essential for all projects. This individual must have many skills. They include the ability to **develop project plans**, **allocate resources**, **resolve conflicts**, identify potential conflicts of interests, promote **safety**, monitor progress against the project plan, report progress, and assure **quality** within traditional project constraints, or what we call the **triangle of balance**.

As illustrated below, the ***triangle of balance*** includes the project scope, time/schedule, and cost/budget..

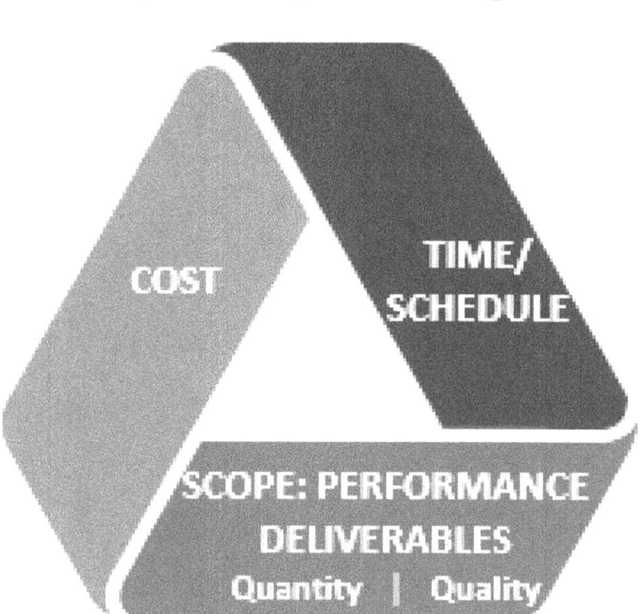

A project manager may have technical knowledge about the project area, and this knowledge may be useful when communicating with various team members. However, a project manager must focus on managing all aspects of the project and not on doing technical work, which is the responsibility of team members.

Managing a project includes monitoring and controlling project tasks and the schedule, identifying problems and leading the team in developing solutions, motivating the team when the going gets tough, and communicating with project stakeholders.

What to do:

☐ For a project you know about, determine the current project phase.

☐ Recognize that project managers must relinquish technical involvement in a project and focus on their project management responsibilities.

☐ Be mindful of traditional project constraints, or what is called the triangle of balance: project scope, time/schedule, and cost/budget.

Other actions:

☐ ___________________________________

☐ ___________________________________

☐ ___________________________________

☐ ___________________________________

☐ ___________________________________

☐ ___________________________________

Remember

✓ The project life cycle is a powerful framework for understanding, organizing, and developing a project. Phases provide an orderly way of envisioning the activities, deliverables, and resource requirements of a project as it progresses in time.

✓ When you work on a project, you will typically find that the phases are recursive. In other words, they go back and forth. For example, the team may have completed the planning phase and be working on executing the project, but then an issue may arise that requires the team to revisit the plan and modify it.

✓ The *triangle of balance* includes the project scope, time/schedule, and cost/budget.

Enhance Your Learning

Watch the following 4-minute video by Jennifer Bridges Whitt. The Project Management Life Cycle.

Whitt, J. (2014). *The project Management Lifecycle..*		Available at: https://www.youtube.com/watch?v=POuGKD3xLqs&t=7s

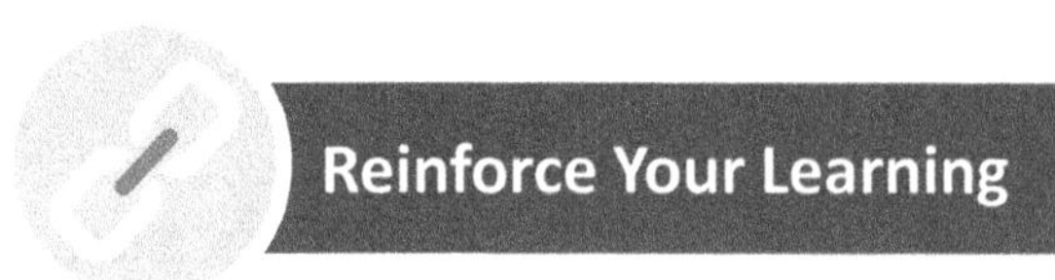

Project Life Cycle Phases. Consider a project that you worked on or observed. Based on your experience, describe the purpose and deliverables of each phase.

Phase	Description and Purpose	Deliverables
Begin the project: **Initiate**		
Plan the project: **Plan**		
Do the project work: **Execute (Monitor and Control)**		
Close the project: **Close**		

Delve Deeper into Project Phases

As previously discussed, after the idea phase, project activities are typically divided into several major phases: initiating, planning, executing, and closing. These phases represent the project life cycle from beginning to end. After an organization vets a project idea and decides to pursue the project, the project Initiate phase begins.

The following is a high-level description of each project phase and associated deliverables. The chart begins with a project idea and then describes the four phases: Initiate, Plan, Execute, and Close, along with the deliverables for each phase.

Be aware that project management is not a linear process. Many project activities will proceed in parallel, and sometimes with overlap and repetition.

Phase	Description/Purpose	Deliverables/Tools
Idea	• Preliminary and general description of a project idea and review of potential benefits.	• Idea appraisal • Decision: yes or no

Phase	Description/Purpose	Deliverables/Tools
Initiate: Define project requirements and scope	• Identify and document business needs and requirements • Define project scope, goals, and objectives • Identify project client, sponsor, and other stakeholders • Develop initial estimates of effort, cost, schedule, and risk • Identify deliverables and acceptance criteria	• Project Description • Project Charter • Feasibility Analysis • Project Scope Statement • Stakeholder Analysis Matrix

Phase	Description/Purpose	Deliverables/Tools
Plan: Plan project from beginning to end	• Define tasks, dependencies, and duration • Refine resource requirements • Develop work plan, milestones, and schedule • Define the project control process, standards, and procedures • Develop a risk management plan • Identify communication needs • Refine and manage expectations	• Schedule Management Plan • Work Breakdown Structure • Status Report Format • Time/Schedule-Tracking Format • Risk Analysis and Management Plan • Stakeholder Communication Matrix • Communication Plan • Project Budget • Benefit-Cost Analysis

Phase	Description/Purpose	Deliverables/Tools
Execute: Do the work while monitoring and controlling the project	• Track actual progress to planned progress • Assess and report status • Coordinate activities and resources • Implement corrective actions and update estimates and project schedule • Manage and support individual and team activities • Manage project priorities • Manage and resolve problems and conflicts • Manage changes to project scope • Assess the quality of deliverables • Communicate work assignments, progress, and issues • Evaluate the performance of individuals and teams	• Updated Project Management Plan Activities • Updated Risk Management Plan • Status Reports • Feedback Forms • Change Order Log • Issues Log • Interim Performance Reviews

Phase	Description/Purpose	Deliverables/Tools
Close: Close and report success level to sponsor and stakeholders	<ul><li>Deliver project results</li><li>Formally end the project</li><li>Evaluate the performance of the project teams, team members, and the project manager</li><li>Evaluate the project, draw conclusions, and document changes and lessons learned</li><li>Identify areas of potential improvement for future projects</li></ul>	<ul><li>Completed Performance Appraisals</li><li>Project Assessment Report</li><li>Updated Project Documents and Records</li><li>Project Deliverables and End Results</li></ul>

For smaller projects, a project manager may decide to combine the initiation and planning phases into a single planning stage. Larger, more complex projects typically keep the definition and planning phases separate, to provide better control of project activities.

Most projects, especially in the definition and planning phases, involve the discovery of new information. If you ignore this information and do not incorporate it into the project life cycle, you will compromise project quality, time, and resources.

Just as important, you will be unable to take advantage of the information. Most projects involve gradually assembling project plans, team members, and organizational resources. When you begin working on a project, you must develop an initial plan. However, this plan is based on available knowledge and sometimes unrealistic expectations. You must expect that some adjustments and modifications to the plan will be needed as you and the team learn more about the project.

What to do:

☐ Vet preliminary project ideas with stakeholders.

☐ Understand that the authorization to pursue a project is provided in a project charter.

☐ Recognize that many project activities proceed in parallel and have some overlap and repetition.

Other actions:

☐ ________________________________

☐ ________________________________

☐ ________________________________

☐ ________________________________

☐ ________________________________

Remember

✓ Most projects, especially in the definition and planning phases, involve the discovery of new information.

✓ Each phase of a project serves a unique purpose. For example, you define the project work in the planning phase.

✓ You will produce specific documents in each phase. For example, a Schedule Management Plan is one of the documents you will develop in the planning phase.

Enhance Your Learning

Watch the following 4-minute video to learn more about the phases of a project by KnowledgeHut .

Knowledgehut. (2019). *4 Stages of Project Life Cycle | Phases of Project Management Life Cycle.*		Available at:. https://www.youtube.com/watch?v=N3N9-RLSbvo

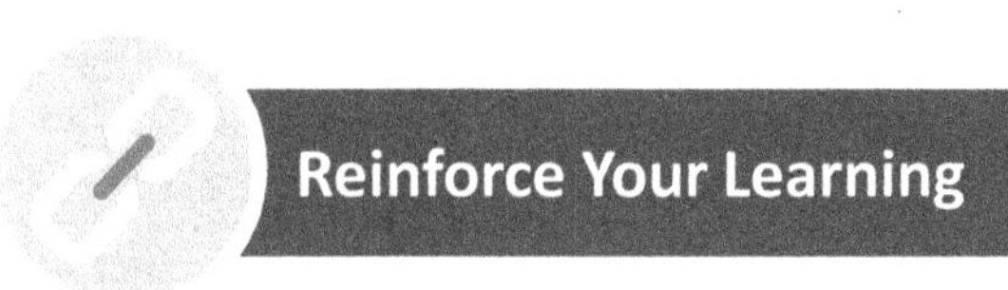

Reinforce Your Learning

Project Phases. Review the phases of the project life cycle and the associated document outputs of each phase. Then list two or three elements you are unsure about or find most challenging and describe why they are challenging.

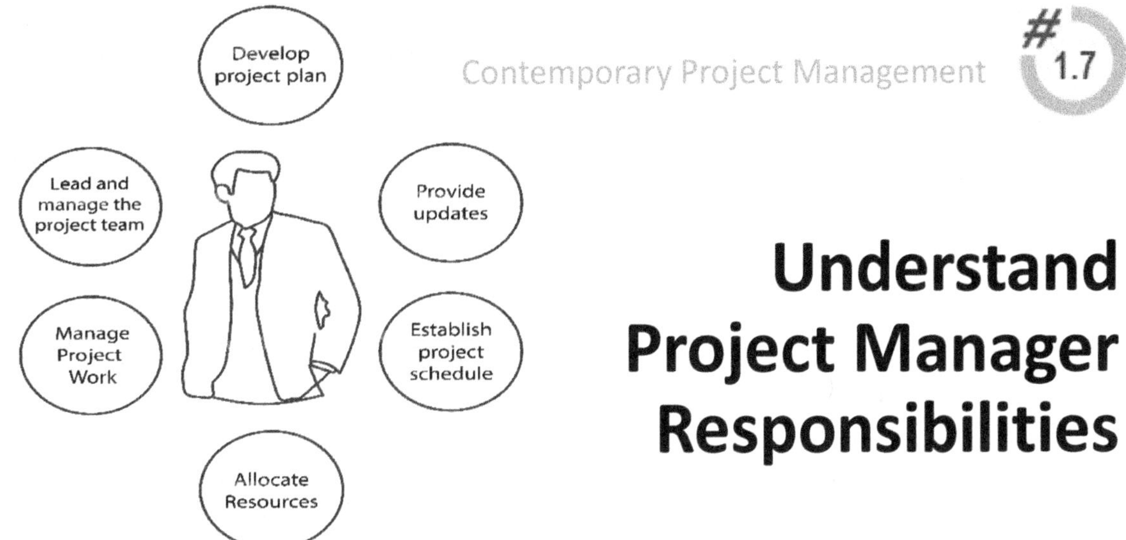

1.7

Understand Project Manager Responsibilities

Project managers are responsible for everything that happens on a project from the time they are assigned until they close the project. On large projects, this can be a long time—months, or even a year or more. Project managers must fulfill these responsibilities while also being available to handle unexpected and urgent situations that threaten project success.

The following is a list of typical responsibilities associated with managing a project. It is not meant to be inclusive, but to show that project managers have a variety of duties.

Project managers are responsible for all these activities throughout the life of a project, though some tend to happen during a particular phase, while others require constant attention. For example, contract negotiation is often early in a project, while communication is a daily job.

Check those areas with which you have experience.

- ☐ Kickoff a project
- ☐ Communicate status and problems and prepare formal project reports
- ☐ Represent/coordinate the project to others in the organization
- ☐ Market the project internally and externally
- ☐ Develop subcontract and joint venture agreement if needed
- ☐ Negotiation and execute contracts
- ☐ Develop a project plan and assign work
- ☐ Allocate resources
- ☐ Manage the project work
- ☐ Manage quality control and quality assurance
- ☐ Resolve conflicts
- ☐ Ensure that billing and collections are done
- ☐ Ensure that project-related personnel policies, particularly those dealing with conduct, conflicts of interest, business confidentiality, and safety are in place and used

Time Management

One of the challenges for project managers is coordinating all the things they must do so that nothing is forgotten, and no deadlines are missed. A time management system can be invaluable by providing a project manager with a way to schedule their work.

Some activities are short-term, whereas others take longer to complete. Daily work, such as making calls and answering questions, requires setting aside time each day. Other activities, for example, conducting meetings and drafting reports, can be scheduled at regular intervals such as weekly or semi-weekly.

Some activities require work over an extended period. For example, on large and complex projects, a project manager may need to set aside time over a month or more to prepare proposals, develop policies and procedures, or do a quality assessment.

What to do:

☐ List your typical daily tasks.

☐ List the weekly and long-term duties you perform.

☐ Investigate time management systems that may help you plan and schedule your work.

Other actions:

☐ _______________________

☐ _______________________

☐ _______________________

☐ _______________________

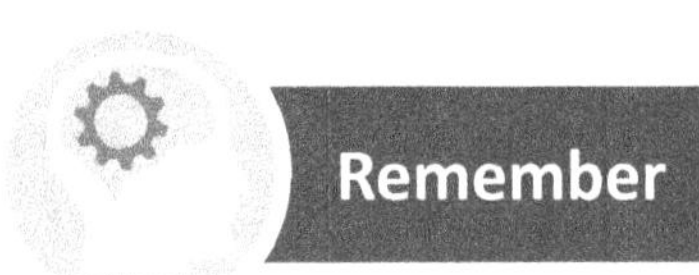

✓ Every project manager has tasks and duties they must fulfill.

✓ These responsibilities occur throughout the life of a project at varying degrees.

✓ Project managers must consider short-term and long-term actions.

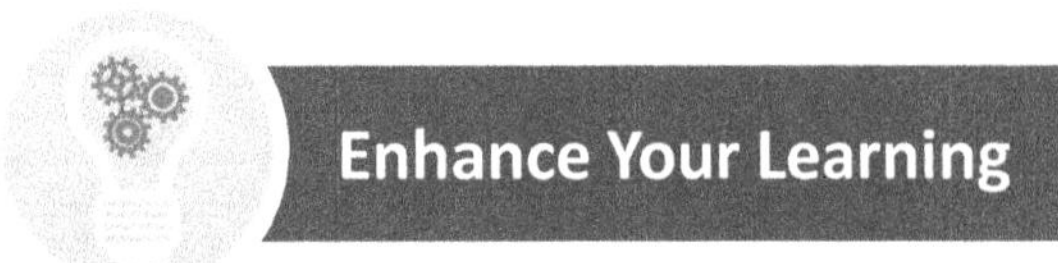

Watch the 7-minute video by Jennifer Bridges Whitt which focuses on daily and weekly job duties for project managers: Top 10 Project Management Responsibilities.

Whitt, J. (2018). *Top 10 Project Management Responsibilities*.		Available at: https://www.youtube.com/watch?v=vMWJyELLSQ4

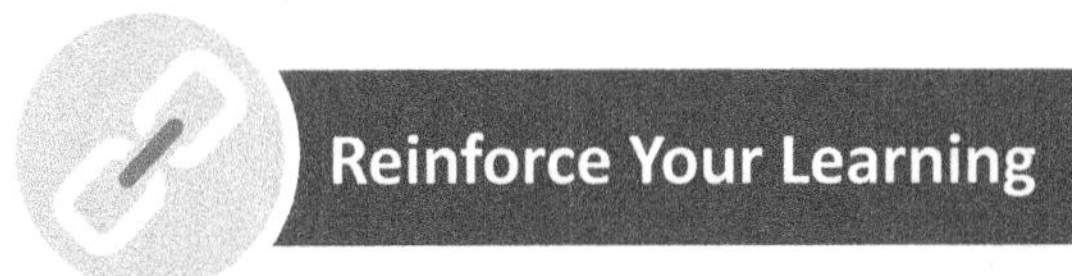

Project Manager Responsibilities. The following are tips for daily project management success developed by Jennifer Bridges, PMP. Based on your experience, can you add to the list?

- Make meetings productive and fun to keep morale and engagement high.

- Make calls personal and intentional, to build relationships and not waste people's time.

- Leave meetings with closure and action items so team members can hit the ground running.

- Resolve issues because a lingering question will not go away by itself, and as the project progresses, it will likely worsen.

- Clarify ambiguity to ensure everyone knows what needs doing, and current project status.

- Ask questions, because by staying curious and engaging your project team, you will learn and lead a better project.

- Keep stakeholders informed. Transparency makes for happier stakeholders. No one wants to feel out of the loop, especially those who are funding the project.

All things are created twice; first mentally then physically. The key to creativity is to begin with the end in mind, with a vision and a blueprint of the desired result.

—Stephen Covey

Consider the Attributes of Successful Project Managers

Effective project managers must do many things well, which means they must have the necessary skills. But they must also have attributes that support these skills.

Attributes are qualities and characteristics that describe a person or thing. They can be vague and difficult to define. For example, caring and being positive are hard to quantify. Although most people know when someone cares or has a can-do attitude, it can be hard to describe.

Project managers must have the same attributes as any successful leader. For example, project managers must:

- Respect the people they work with and treat them accordingly.
- Understand people and how they function in teams.
- Appreciate the skills of the team members and trust them to do their jobs.
- Care what happens on the project and be willing to address problems that threaten it.
- Work diligently to ensure the project is completed according to plan.
- Care about the organization's goals.

Over time, successful project managers develop such attributes—respect, understanding, appreciation, trust, caring, courage to confront problems, and a strong work ethic—by building on their strengths and acknowledging and working to overcome their weaknesses.

The following are some of the behaviors expected from project managers. As you review the list, consider what qualities or characteristics a project manager might need to be effective and check those items for which you would like to build your competence.

Successful project managers know how to:

- ☐ Sacrifice technical involvement for management responsibilities
- ☐ Take responsibility for what team members do, both good and bad
- ☐ Sacrifice ego and promote team involvement and recognition
- ☐ Trust the skills of the team members
- ☐ Coordinate and work comfortably with various technical disciplines
- ☐ Share information with team members and stakeholders
- ☐ Keep stakeholders informed
- ☐ Delegate responsibility and authority while keeping accountability
- ☐ Work with and through others, rather than as an individual contributor

    ☐ Safeguard against allowing personal feelings or biases to interfere with managing the project

    ☐ Help the team, even when doing so conflicts with the project manager's own work and obligations

What to do:

☐ Identify the qualities and characteristics you typically display in a project.

☐ Identify your attributes that are strengths and build on them.

☐ Determine what attributes you could improve and find resources to help you grow.

Other actions:

☐ _______________________________

☐ _______________________________

☐ _______________________________

☐ _______________________________

☐ _______________________________

✓ A project manager must do many things well.

✓ Attributes are the qualities and characteristics someone or something has, for example, caring or having a can-do attitude.

✓ Project managers must have realistic views of their strengths and weaknesses.

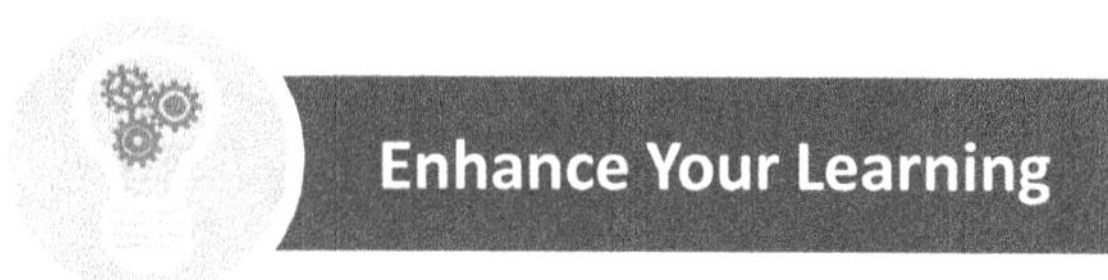

Review the following article by Nutcache (2022), to learn ten attributes of an effective project manager.

Nutcache. (2022). *10 Attributes of an Effective Project Manager.*  Available at: https://www.nutcache.com/blog/10-attributes-effective-project-manager/

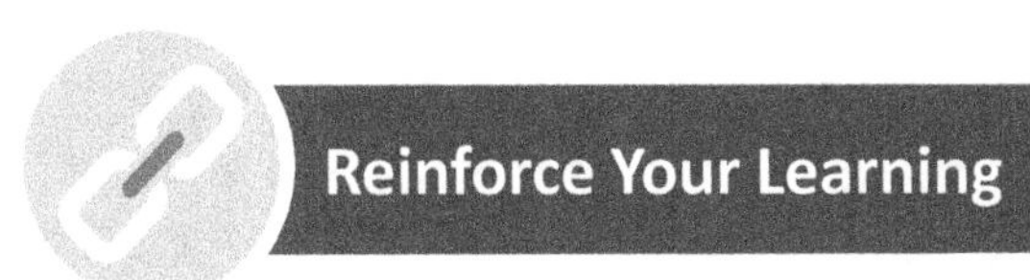

Project Manager Attributes. Based on your experience, which of the following behaviors are your strengths, and which are your weaknesses? Write an "S" in the space provided to indicate a strength and a "W" to indicate a weakness. Then jot down several actions you might take to overcome the weaknesses,

______ 1. Effective communication skills

______ 2. Leadership and delegation skills

______ 3. Good problem solving and decision-making skills

______ 4. Proven technical expertise/skills

______ 5. Good team builder

______ 6. Confident and cool under pressure

______ 7. Good influence and negotiation skills

46

Plans are only good intentions unless they immediately degenerate into hard work.

—Peter Drucker

Realize the Diverse Roles Project Managers Fulfill

Project managers must fulfill several functional roles as the project progresses and moves through its life cycle.

Below are descriptions of the critical roles project managers satisfy. Various roles will be more prominent at one time or another depending on what part of the project life cycle a team is working on, the problems or challenges the project faces, and what may be required to bring the project to successful completion.

For example, a project manager may fulfill a visionary role when the project is first being considered, and more of a problem analyzer role when the project is being worked on and problems arise.

Communicator

A communicator is skilled in sending and receiving messages and information in a variety of situations and media, such as presentation, interpersonal relationships, and writing. In this role, the project manager must communicate stakeholder needs to the team and communicate team progress and issues to stakeholders.

Visionary

A visionary can foresee products, services, and results not yet apparent to others. The visionary is not psychic, but they listen and gather information and work to assess how things are going. They pay attention and communicate what they see. This role requires a "big-picture person," one who sees opportunities and possibilities and who can help others see them as well.

Supporter

A supporter takes ownership of project work and deliverables and encourages others to do likewise. A supporter promotes the project internally and externally.

Motivator

A motivator helps create a positive climate and has a positive impact on work performance.

Planner

A planner helps define and structure the steps and tasks required to complete a project successfully.

Organizer and Resource Coordinator

This role plans for resources, ensuring that the right resources get to the right places at the right time and in the right quantities.

Team Leader

A team leader is responsible for the day-to-day activities of the project team and has a crucial role in influencing team members, stakeholders, and project outcomes in a positive way.

Problem Analyzer and Decision Maker

A problem analyzer identifies problems and determines their causes, brainstorms, advances possible solutions for team discussion, and recommends fixes in accordance with project goals.

Facilitator

A facilitator helps implement the project vision, is service-minded, and can focus on the big picture while respecting the contribution of each team member and supporting the team in achieving their goals.

What to do:	**Other actions:**
☐ Identify the roles you typically fulfill in projects.	☐ _______________________
☐ Identify the roles that are your strengths and build on them.	☐ _______________________ ☐ _______________________
☐ Determine what roles you could improve and find resources that could help you develop your skills.	☐ _______________________

✓ Project managers play many roles throughout the life cycle of a project.

✓ Peers can serve as free career development resources.

✓ You are responsible for your career and professional development.

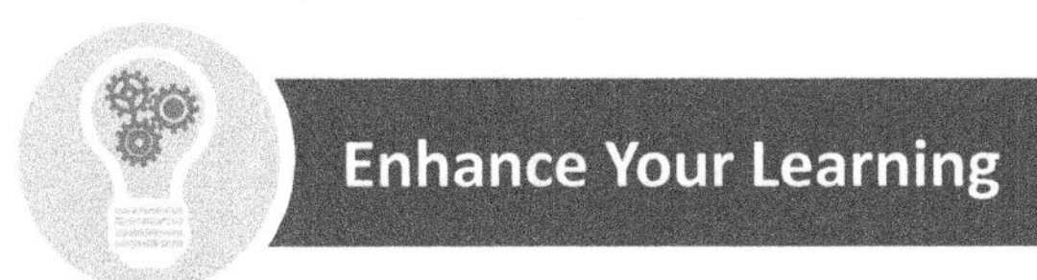

Watch the following 17-minute video by Jason Dodd of PM Perspective (2014), to learn more about the role of the project manager by PM Perspective.

Dodd, J. (2014). *The Role of The Project Manager.*

Available at:
https://www.youtube.com/watch?v=dUhJuB69ZBo

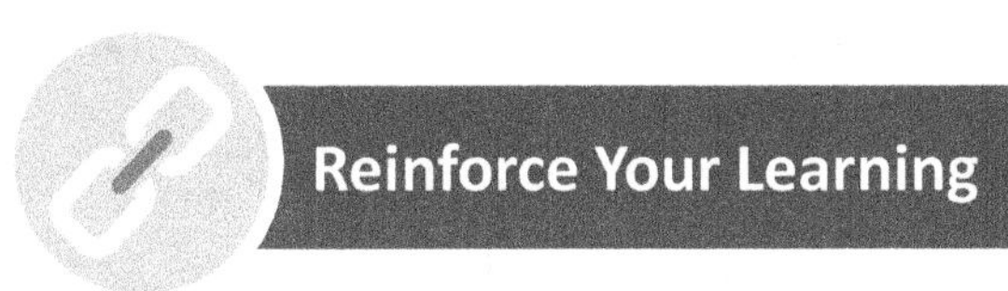

Project Manager Roles. Based on your experience, check the items you feel are essential. Can you add anything?

A communicator:

- ❑ Conducts project reviews and walk-throughs.
- ❑ Does not interpret "no news" as "good news."
- ❑ Presents and reports periodically according to a schedule and as requested.
- ❑ Ensures that all reports and presentations are clear and well organized.
- ❑ Uses specific and clear terminology when communicating and no buzz words.
- ❑ Meets with others throughout the project to ensure a full understanding of stakeholder needs and concerns.
- ❑ Follows up with stakeholders after the project to ensure their satisfaction.
- ❑ Identifies all groups that may be affected by the project and seeks their input.
- ❑ Holds regularly scheduled and productive meetings with the project team and ensures representation from all areas affected by the project. Discusses status, resolves issues, and shares information.
- ❑ Informs management and stakeholders promptly when problems arise.
- ❑ Speaks on behalf of management and stakeholders in project meetings.
- ❑ Other:

A visionary:

- ❑ Possesses a business orientation.
- ❑ Considers project goals in a cross-functional manner.
- ❑ Thinks not only in concrete terms but in a conceptual context.
- ❑ Reviews the overall picture.
- ❑ Stays ahead of the project team.
- ❑ Develops a preliminary study with the project team to identify business opportunities, project requirements, project scope, and project benefits.

- ❑ Works closely with stakeholders to define project objectives and key results/goals.
- ❑ Acts today to prevent tomorrow's problems and addresses future requirements.
- ❑ Communicates factors necessary for success.
- ❑ Views the project not just in the present but also in the context of how business and technology will perform in the future.
- ❑ Anticipates and plans for the impact of the project on related systems and areas.
- ❑ Considers both short- and long-term implications of the project.
- ❑ Ensures a common understanding of the project objectives by all stakeholders and team members.
- ❑ Ensures a common understanding of the project scope by all stakeholders and team members.
- ❑ Other:

A supporter:

- ❑ Uses relationships with people from different areas in the organization to resolve issues and help guide team members.
- ❑ Assumes ownership of the project.
- ❑ Publicizes and markets the project internally and externally.
- ❑ Builds business ownership by seeking extensive business participation during the initiating and planning phases of a project.
- ❑ Encourages stakeholders and team members to assume ownership.
- ❑ Empowers others to achieve project objectives.
- ❑ Represents the stakeholder and team perspectives in the company.
- ❑ Other:

A motivator:

- ❑ Manages performance issues.
- ❑ Keeps an open-door policy and wants to hear ideas and problems.
- ❑ Positively recognizes individual and team accomplishments and results.
- ❑ Exerts whatever effort is needed to achieve project objectives.
- ❑ Manages expectations by ensuring that what the team promises gets delivered.
- ❑ Provides rewards aimed at the needs and interests of team members.
- ❑ Identifies with project quality and success.
- ❑ Rewards and recognizes people as milestones are achieved.
- ❑ Presents a positive image to set the tone for the project team.
- ❑ Other:

A planner:

- ❑ Develops and proposes strategies.
- ❑ Develops the project management plan and determines assignments.
- ❑ Establishes lines of responsibility and accountability with stakeholders.
- ❑ Understands business needs/requirements, time, and cost pressures.
- ❑ Balances ideal technical solutions and project scope against business deadlines and priorities.
- ❑ Sets and supports high standards of quality for all team members.
- ❑ Breaks a large or complex project into meaningful sub-projects.
- ❑ Develops a project plan that includes resource needs, budget, and time. Involves project team members in the detailed planning of the project.
- ❑ Maintains a detailed baseline plan, showing what to do and by what time.
- ❑ Establishes priorities.
- ❑ Sets objectives in cooperation with the project team.
- ❑ Assesses risks and develops an effective risk management plan.
- ❑ Other:

A team leader:

- ☐ Works to ensure stakeholder satisfaction.
- ☐ Models appropriate behavior for team members.
- ☐ Facilitates project work and project meetings.
- ☐ Leads the project team and stakeholders to effectively accomplish project objectives and adjusts as situations change.
- ☐ Networks with others, both inside and outside the organization.
- ☐ Builds a capable team.
- ☐ Provides direction for the project by influence and not force.
- ☐ Makes tough decisions after gathering all the necessary input and facts.
- ☐ Adapts to situations as they change.
- ☐ Keeps learning and growing and is self-confident.
- ☐ Shows flexibility and is willing to change when appropriate.
- ☐ Brings out the best in all team members.
- ☐ Defines clear roles and performance expectations for team members.
- ☐ Is objective when dealing with conflict.
- ☐ Focuses team members on project objectives and deliverables.
- ☐ Identifies schedule and budget/cost issues of requested scope changes and communicates them to stakeholders.
- ☐ Accepts responsibility for resolving project issues promptly.
- ☐ Works to influence stakeholders and team members in a positive way.
- ☐ Maintains accountability throughout the project and holds others accountable for meeting their goals.
- ☐ Other:

A problem analyzer and decision-maker:

- ☐ Determines what data is essential and analyzes it efficiently.
- ☐ Sets priorities to ensure the project is completed within budget, on time, and within quality standards.
- ☐ Focuses on problems and solutions, not on who is at fault.
- ☐ Cultivates the proper atmosphere for practical problem-solving.
- ☐ Emphasizes quality in all decisions.
- ☐ Makes tough decisions without procrastinating.
- ☐ Identifies problems early enough so they can be dealt with before they develop into time-consuming issues.
- ☐ Works with financial staff to manage the project's costs and budget.
- ☐ Develops innovative and creative approaches to problems.
- ☐ Takes calculated and prudent risks.
- ☐ Takes corrective action to keep the project on schedule and within budget.
- ☐ Confronts problems with others in a timely and direct manner.
- ☐ Other:

A facilitator:

- ❏ Delivers services and products to stakeholders successfully.
- ❏ Shows skill and is comfortable in dealing with ambiguity.
- ❏ Acts to overcome obstacles and limitations.
- ❏ Develops an atmosphere where all team members are comfortable and involved in the project openly and honestly.
- ❏ Considers the implications of finances on the project and recognizes the significance of return on investment.
- ❏ Shows awareness of the time required by team members to do jobs.
- ❏ Provides team members with assignments and training to support their growth.
- ❏ Gives direct, specific, constructive, and timely feedback to team members.
- ❏ Other:

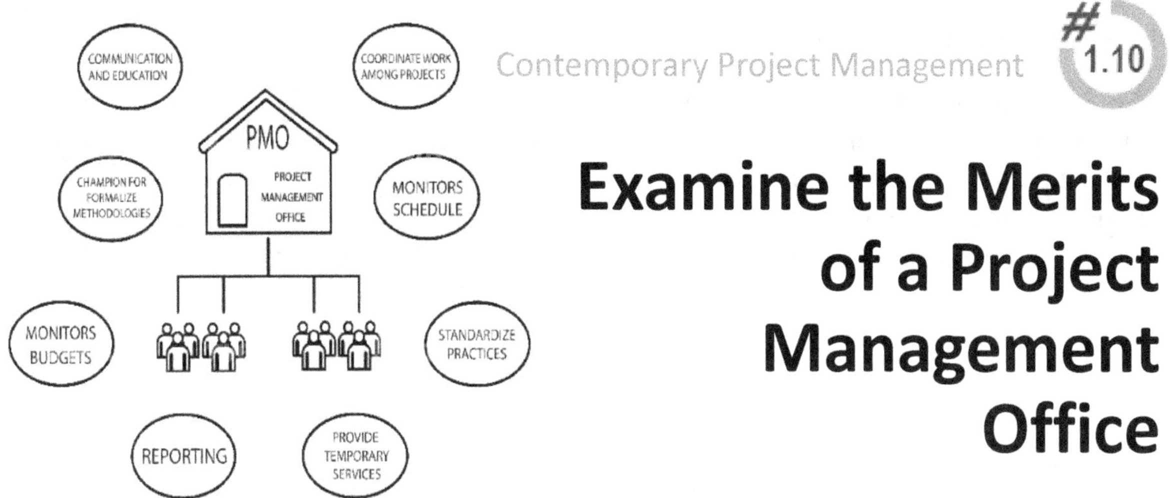

Examine the Merits of a Project Management Office

When you manage a project, you will find it helpful to understand how your organization is structured. Organizations may be designed or structured differently to manage projects.

Traditional Functional Organizational Structure

In the traditional organization, every group of related activities is assigned to a single functional manager and department. Each department head reports to a higher-level manager. Each department employs people who have similar skills and resources.

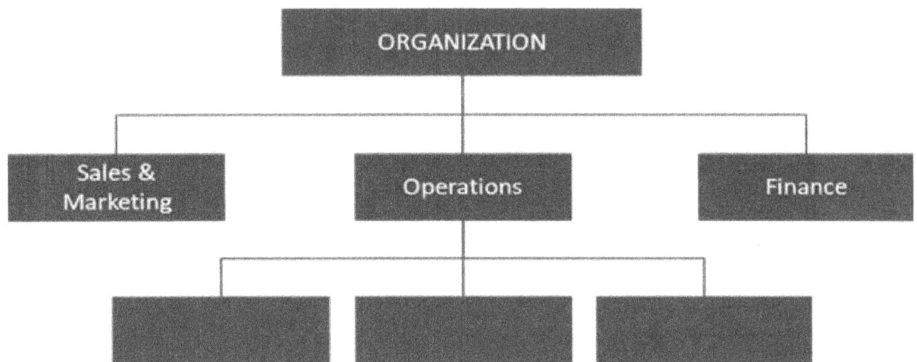

A **major drawback** for functional organizations in dealing with projects is cross-departmental communication and coordination presents challenges. Therefore, many projects are limited to the boundaries of the functional department.

A typical line and staff organization builds on the idea of the functional organization. The line consists of those employees who produce the organization's products and services. The staff consists of those who support the line employees such as accounting, maintenance, human resources, and the like.

In the line and staff organization, the chain of command is clear. While the chain of command can provide order and control, it may work against the needs of those who manage cross-functional projects. Project managers in line and staff organizations may experience challenges because they need to use a variety of types of power and influence to get their work done effectively.

Matrix Organizational Structure

The matrix organization is less traditional and is set up to do projects. For example, in an engineering firm, the designated project managers—or project coordinators—call upon the talents of employees from functional departments—or disciplines such as mechanical or electrical engineering—for a designated time period. After the project is completed, they return to their designated units until they are needed for another project.

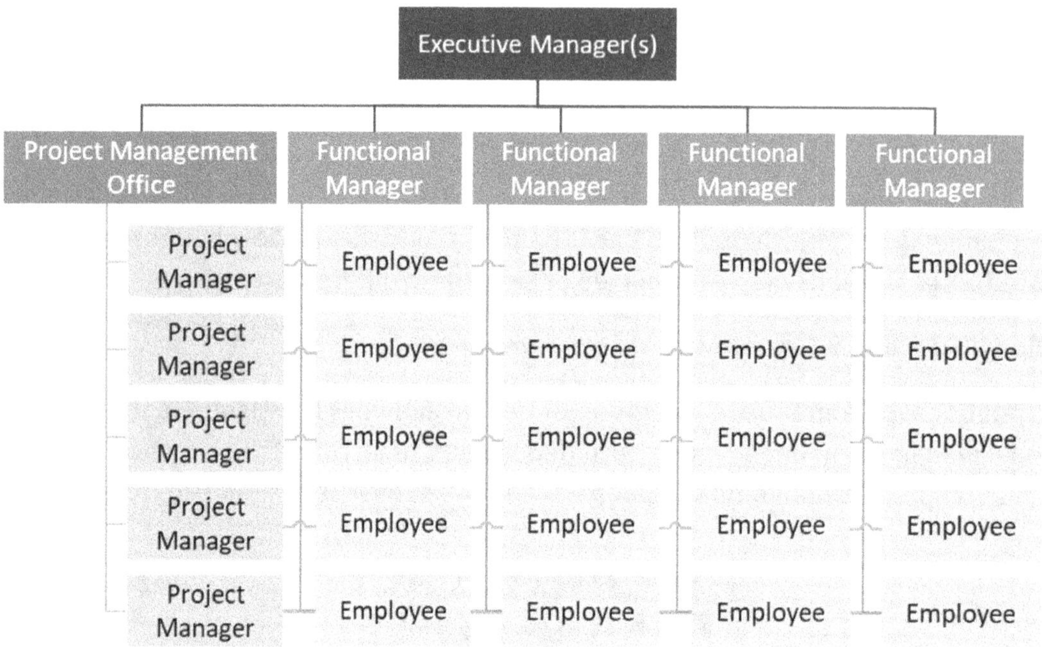

In matrix firms without a project management office, the project managers or project coordinators are often selected from the various functional areas and serve a dual role where the chain of command is less clear. There are frequently multiple reporting relationships.

Projectized Organizational Structure

Organizations can enhance their project management process by adopting organization-wide methods. The advantages of doing this are many. It comes down to the basic idea of not expending resources to reinvent the wheel. For example, when people move from one project to another, they would not need training on project procedures, and from one project to the next, people in the organization would be familiar with how projects are reported.

The projectized organization is often used on larger and longer-term projects where the project manager has a great deal of authority over the projects and over the project team members. Many organizations have segmented their project management activities under a discrete **Project Management Office** (PMO) with its own project executive.

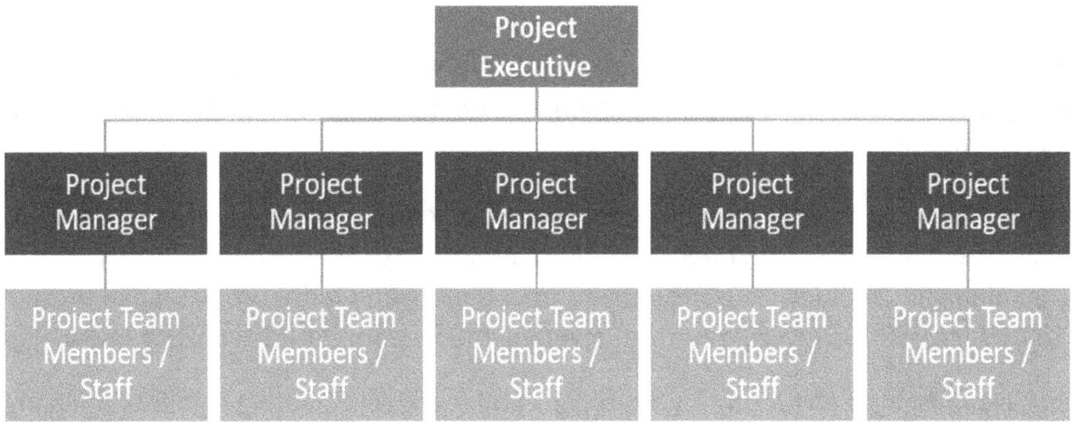

In a projectized organization all the team members work for the project manager who has control over the resources. Typically, the project manager has close ties to the client and can accommodate changes quickly. Team members are usually assigned full-time to the project and often co-located. A centralized **project management office** (PMO) can facilitate organization-wide project management in a variety of ways. For example, the PMO in an organization can:

- Coordinate work among projects

- Standardize project management practices and methods

- Conduct project management training

- Provide temporary project management services

- Champion the value of a formalized project management methodology throughout the organization

A PMO promotes consistent project management practices and methods and the use of standard project management tools; assures standardization in project execution and delivery; provides a resource for project management expertise; and coordinates multinational project management knowledge (Kerzner, 2017).

There is no single way of doing every project. By definition, each project is unique. However, a standard methodology for how to conduct projects can be established. For example, the recognized phases of the project life cycle—initiate, plan, execute, perform work while monitoring and controlling it, and close—are common to all projects. This general framework gives everyone involved in a project a roadmap to follow, even though the specific activities within each phase may not all be the same from one project to another.

Another area of commonality among projects is the **standardization** of processes for project execution and delivery. For example, a PMO can help establish practices that might include a consistent set of criteria for project approval, determination of what project management software to use, a consistent set of success criteria, and how and when a project can be closed.

Another way a PMO can serve an organization is by providing **subject matter expertise**. Having a PMO staffed with people who have project management experience in a variety of environments provides a significant resource for an organization. A pool of project management professionals can be instrumental in enabling an organization to consistently deliver successful projects.

Coordinating project knowledge can be cost-effective in other ways. For example, there once may have been little need to understand the complexities of conducting a global project, but that has changed. In today's business environment, more and more projects span the globe, and to be successful, organizations must leverage their comprehensive project knowledge.

Besides establishing a PMO, other ways organizations are coordinating and leveraging project best practices include the use of *lessons learned*, which is a way to document how things could have been done better after a project is complete, and *enterprise knowledge databases where* information about past projects is stored for use by future teams.

What to do:	**Other actions:**
☐ Visit a PMO or read an article about one to learn how a real-world PMO operates.	☐ _________________________
☐ Search online for two examples of a PMO. Does anything you learn apply to your environment?	☐ _________________________
	☐ _________________________
☐ Consider the services a PMO could provide. How could your organization benefit from having one?	☐ _________________________

Remember

✓ The concept of a PMO is a centralized collection of project management expertise.

✓ A PMO promotes the maturation of project management best practices and methods within an organization.

✓ Having a PMO can improve the likelihood of projects being completed successfully.

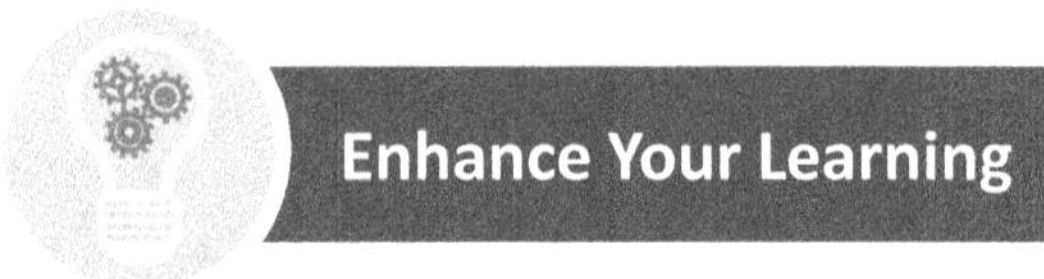

Enhance Your Learning

View the 4-minute video which describes different organizational structures employed for managing projects.

Kumar, K. (2016). *Organization structure Influence project management | Functional Matrix Projectized organization.*		Available at: https://www.youtube.com/watch?v=pbtqqZm8Ak0

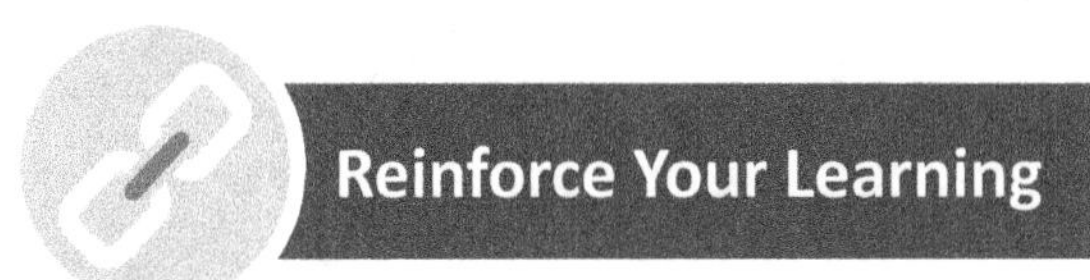

Project Management Office (PMO). Consider the following case study of a project within an organization that did not have a PMO.

Pipeline Education, Inc. has been growing over the last two years. They have added new employees and are struggling to keep up with requests for new online courses. Pipeline promoted one of their technical staff, Diane, who knew a lot about online education, to be a project manager. However, she had never taught a course and was not familiar with how to plan and run a project. The other project managers did not have time to train Diane or assist her. She did the best she could with her limited exposure to project management methods.

In a couple of months, Diane's project fell behind schedule. Then the organization discovered that the work Diane's team completed so far had quality issues. The software tools were not the best for online learning, the courses did not meet ADA requirements, the team's analytical tools to measure learning outcomes had deficiencies, and the methods to plan and schedule work did not meet the organization's standards.

Below, identify and describe at least two or three ways a PMO could have helped address the problems in Diane's project.

58

Project management can be defined as a way of developing structure in a complex project, where the independent variables of time, cost, resources and human behavior come together.

—Rory Burke

Summary

A Final Word: Contemporary Project Management

You have completed this chapter on Contemporary Project Management. You were introduced to the Project Management Institute (PMI)®, their certifications, and A Guide to the Project Management Body of Knowledge (*PMBOK*)®.

You were introduced to the characteristics of a project, the life cycle of a project, and the project management process. You also had a brief look at responsibilities, attributes, and diverse roles of a project manager as well as the potential merits of a project management office.

Information you reviewed in the chapter include:
- Distinguishing project management from conventional management
- Describing the phases of a project from beginning to end
- Understanding the benefits of using a project management methodology
- Recognizing the attributes successful project managers bring to their projects
- Being aware of the critical roles project managers fulfill

In the next *Mastering Project Management: Planning for Performance* chapter, we will focus on Defining Project Scope.

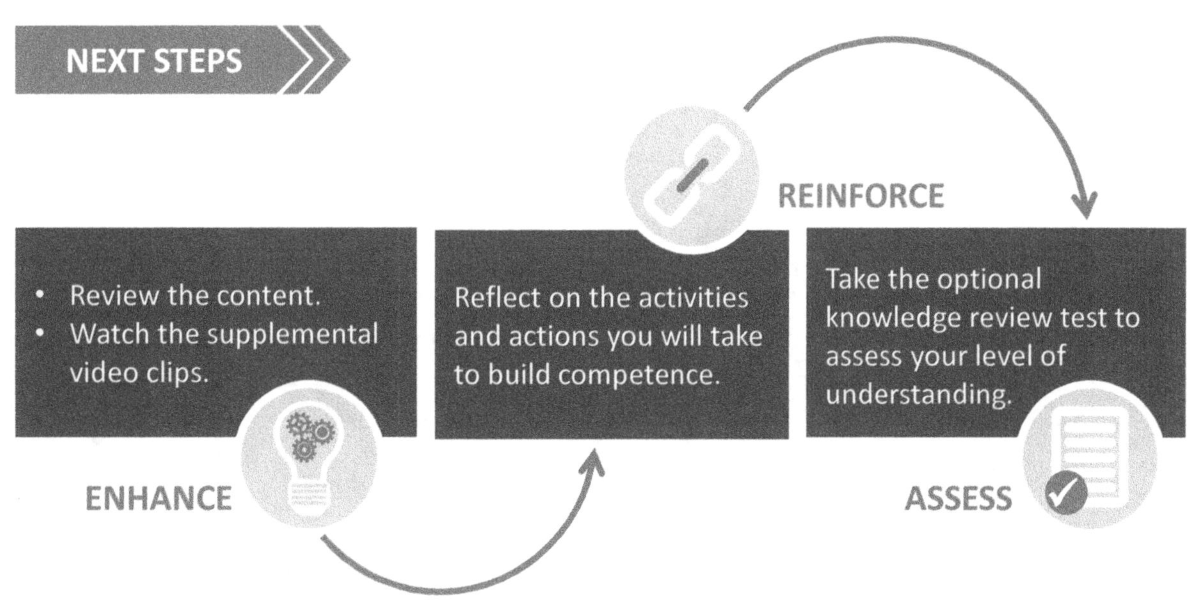

Notes:

<table>
<tr><td colspan="1">

Contemporary Project Management—Recap Checklist

This course describes project management and its benefits, reviews the project management life cycle, describes the functions a project manager performs, and examines the challenges that affect most projects at one time or another with a focus on the human side.

</td></tr>
</table>

1. Define Project and Project Management

- ☐ Review the Project Management Institute (PMI) website (www.pmi.org).
- ☐ Identify some projects you have been involved in or observed.
- ☐ Brainstorm issues, problems, and training needs related to those projects.

2. Become Familiar with the Project Management Institute (PMI)

- ☐ Discuss project management certifications with your manager for applicability.
- ☐ Find a certified associate and ask them about the advantages of being certified.
     Pursue one of the certifications, if appropriate.

3. Know the Benefits of Using a Project Management Methodology

- ☐ Consider the benefits of following a step-by-step planning process.
- ☐ Identify the sources/kinds of information you need access to.
- ☐ Reflect on the risks in your project and what their impact could be.

4. Be Aware of How Projects Begin, the People Involved, and Quality

- ☐ Identify the criteria your organization uses to judge project success.
- ☐ Define what quality means for projects you (or your organization) have completed.
- ☐ Complete an Idea Appraisal Worksheet for a project that you or someone in your organization is considering.

5. Explore the Project Life Cycle

- ☐ For a project you know about, determine the current project phase.
- ☐ Recognize that project managers must relinquish technical involvement in a project and focus on their project management responsibilities.
- ☐ Be mindful of traditional project constraints or what is called the *triangle of balance*: scope, time/schedule, and cost/budget.

6. Delve Deeper into Project Phases

- ☐ Vet preliminary project ideas with stakeholders.
- ☐ Understand that the authorization to pursue a project is provided in a project charter.
- ☐ Recognize that many project activities proceed in parallel and have some overlap and repetition.

7. Understand Project Manager Responsibilities

- ☐ List your typical daily tasks.
- ☐ List the weekly and long-term duties you perform.
- ☐ Investigate time management systems that may help you plan and schedule your work.

8. Consider the Attributes of Successful Project Managers

- ☐ Identify the qualities and characteristics you typically display in a project.
- ☐ Identify your attributes that are strengths and build on them.
- ☐ Determine what attributes you could improve and find resources that could help you grow.

9. Realize the Diverse Roles Project Managers Fulfill

- ☐ Identify the roles you typically fulfill in projects.
- ☐ Identify the roles that are your strengths and build on them.
- ☐ Determine what roles you could improve and find resources that could help you develop your skills.

10. Examine the Merits of a Project Management Office

- ☐ Visit a PMO or read an article about one to learn how a real-world PMO operates.
- ☐ Search online for two examples of a PMO. Does anything you learn apply to your environment?
- ☐ Consider the services a PMO could provide. How could your organization benefit from having one?

Action Planning | **Competency #1** >>

Planning and Evaluation

Establishes policies, guidelines, plans, and priorities; plans and coordinates with others; aligns required resources; monitors progress and evaluates outcomes; improves organizational efficiency and effectiveness.

Briefly Describe how improvement in this competency will help you achieve important results or better meet your job responsibilities.

List courses, books, and independent study opportunities that could help you develop this competency.

Action Planning | **Competency #1** >>

Identify one or more people who could help you, either as a role model or source of
information. Write any questions you want to ask each person

What specific steps will you take?	Start Date	Finished

Resource Management

Demonstrates awareness of technical resources; knows how to apply resources to achieve desired outcomes.

Briefly describe how improvement in this competency will help you achieve important results or better meet your job responsibilities.

List courses, books, and independent study opportunities that could help you develop this competency.

Identify one or more people who could help you, either as a role model or source of information. Write any questions you want to ask each person

What specific steps will you take?   **Start Date**   **Finished**

Leadership and Coaching

Models and encourages high standards of ethical behavior; adapts leadership styles to situations and people; empowers, motivates, guides, and coaches Planning and Evaluation.

Briefly describe how improvement in this competency will help you achieve important results or better meet your job responsibilities.

List courses, books, and independent study opportunities that could help you develop this competency.

Identify one or more people who could help you, either as a role model or source of information. Write any questions you want to ask each person

What specific steps will you take? **Start Date** **Finished**

68

If everything seems under control, you're not going fast enough.

—Mario Andretti

Contemporary Project Management

Knowledge Review Test

Part A. Knowledge Review Test—Questions

Part B. Knowledge Review Test—Answer Sheet

Part A. Knowledge Review Test - Questions

Contemporary Project Management

1. Which is not a characteristic of a project?
    A. Temporary.
    B. Definite beginning and end.
    C. Interrelated activities.
    D. Repeats itself every month.

2. Wes is a new project manager who has never managed a project before and has been selected to plan a new project. It would be best in this situation to rely on _______ during the planning to improve his chance of success?
    A. his intuition
    B. his training
    C. historical records
    D. responsibility charts

3. The Project Management Institute was founded in what year?
    A. 1941
    B. 1939
    C. 1969
    D. 1991
    E. 2001

4. What certification designation offered by the Project Management Institute (PMI) shows the holder has the skills and knowledge to manage projects from beginning to end?
    A. Certified Associate Project Manager (CAPM®)
    B. Program Management Professional (PgMP®)
    C. Agile Certified Practitioner (PMI-ACP®)
    D. Project Management Professional (PMP)®

5. For a project to be successful, which of the following criteria must be met?
    A. Scope requirements.
    B. Client satisfaction.
    C. Project schedule.
    D. Completed on budget.
    E. All the answers are correct.

6. The CEO of your organization asked you to work on a variety of strategic business activities aimed at building, sustaining, and advancing the initiatives across your organization. What does this best describe?
   A. Portfolio analysis of the organization.
   B. Project management information system.
   C. System for value delivery.
   D. Agile and Scrum analysis.

7. Organizations review ideas for projects in many ways, but most include:
   A. A description of the project.
   B. An assessment of the benefits.
   C. A comparison against other projects.
   D. A sponsor.
   E. All the answers are correct.

8. Your project should provide value to your customers. Which of the following does not describe value?
   A. The ability to use unique features or functions of a product.
   B. The anticipated use of scrap caused by selecting an easy solution instead of a better solution that would take longer.
   C. Financial metrics, such as the benefits less the cost of achieving those benefits.
   D. The worth, importance, or usefulness of something.

9. Which of the following is not a stage in the project management cycle?
   A. Defining project scope.
   B. Planning the project.
   C. Allocating project resources.
   D. Implementing, monitoring, and controlling.
   E. Closing the project.

10. The final result or consequence of a process or project that focuses on the benefits and value that the project was undertaken to deliver is known as:
   A. Outcome.
   B. Value.
   C. Deliverable.
   D. Product.

11. Project integration means:
   A. Familiarizing team members with the project.
   B. Assigning team members to teams.
   C. Putting all the pieces of a project into a cohesive whole.
   D. Putting the pieces into a program.

12. When the project manager delivers a project charter, it should be in which phase or stage of project management?

    A. Planning the project.
    B. Implementing, monitoring, controlling.
    C. Initiation or beginning of the project.
    D. Closing the project.

13. Project managers do not address which of the following project management activities?

    A. Planning the project.
    B. Executing the work.
    C. Tracking programs.
    D. Documenting performance.
    E. None of the answers are correct.

14. A project stakeholder may include:

    A. End users.
    B. Suppliers.
    C. Citizens.
    D. All the answers are correct.

15. An effective project manager possesses a wide variety of skills and attributes. Which of the following statements does not set a good example of a Project Manager Attribute?

    A. Delegates responsibility, but not accountability.
    B. Sticks to his own work and obligations and encourages each team member to do the same.
    C. Shares information with team members.
    D. Is comfortable coordinating activities in various technical disciplines.

16. Projects have the least attention in what form of organization?

    A. Functional.
    B. Matrix.
    C. Expeditor.
    D. Coordinator.

17. One of the main advantages of a matrix organization is:

    A. Improved project manager control over resources.
    B. More than one boss for project teams.
    C. Communication is easier.
    D. Reporting is easier.

18. Which of the following is not considered a key role of the project manager?
    A. Motivator.
    B. Team Leader.
    C. Reviewer.
    D. Facilitator.

19. Which of the following helps with coordinating efforts between projects, standardizing the project management methodology, conducting project management training, providing temporary project management services internally, and championing the value of formalized project management methodology throughout an organization?
    A. Your project sponsors.
    B. A project management office (PMO).
    C. Your stakeholders.
    D. The Project Management Institute (PMI.)

20. Wes has little project experience, but he has been assigned as the project coordinator of a several new projects. Because he will be working in a matrix organization to complete his projects, he can expect communication to be:
    A. Simple .
    B. Open and honest.
    C. Complex.
    D. Hard to automate.

Part B. Knowledge Review Test—Answer Sheet

1.   ___ A   ___ B   ___ C   _X_ D   ___ E

2.   ___ A   ___ B   _X_ C   ___ D   ___ E

3.   ___ A   ___ B   _X_ C   ___ D   ___ E

4.   ___ A   ___ B   ___ C   _X_ D   ___ E

5.   ___ A   ___ B   ___ C   ___ D   _X_ E

6.   ___ A   ___ B   _X_ C   ___ D   ___ E

7.   ___ A   ___ B   ___ C   ___ D   _X_ E

8.   ___ A   _X_ B   ___ C   ___ D   ___ E

9.   ___ A   ___ B   _X_ C   ___ D   ___ E

10.   _X_ A   ___ B   ___ C   ___ D   ___ E

11.   ___ A   ___ B   _X_ C   ___ D   ___ E

12.   ___ A   ___ B   _X_ C   ___ D   ___ E

13.   ___ A   ___ B   ___ C   ___ D   _X_ E

14.   ___ A   ___ B   ___ C   _X_ D   ___ E

15.   ___ A   _X_ B   ___ C   ___ D   ___ E

16.   _X_ A   ___ B   ___ C   ___ D   ___ E

17.   _X_ A   ___ B   ___ C   ___ D   ___ E

18.   ___ A   ___ B   _X_ C   ___ D   ___ E

19.   ___ A   _X_ B   ___ C   ___ D   ___ E

20.   ___ A   ___ B   _X_ C   ___ D   ___ E

Defining Project Scope

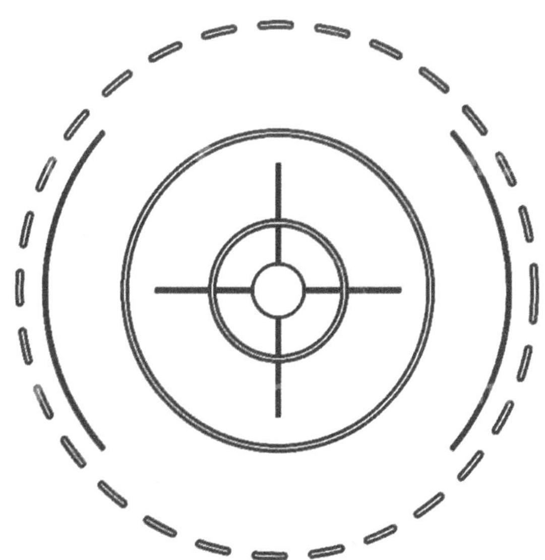

76

Obstacles are those frightful things you see when you take your eyes off your goal.

—Henry Ford

Overview

Defining Project Scope

"Do not repeat the tactics which have gained you one victory, but let your methods be regulated by the infinite variety of circumstances."

—Sun Tzu c. 490 BC

When it comes to completing projects, what you do not know can hurt you and your organization. Doing the work might be 90 percent of the job, but project planning will impact the entire project and help ensure success. A project well done is always a project well planned.

The topics in this chapter introduce you to many of the issues involved with project planning and discuss how to avoid project pitfalls.

In this chapter, you will learn how to:

- Conduct a stakeholder analysis
- Define the project scope
- Recognize common causes of project failure
- Evaluate the critical factors associated with starting projects and your readiness
- Develop a Work Breakdown Structure (WBS)

The competencies associated with this chapter include:

Planning and Evaluation **Resource Management** **Communication**

Take Your Temperature for Defining Project Scope

The following assessment will help you identify what you know about defining the scope of a project and the areas where you can improve. With you and your organization in mind, read each statement below, and consider how much you agree with it. Use the numbers from 1 to 10, where **1** means you **strongly disagree** and **10** means you **strongly agree**.

DISAGREE   1   2   3   4   5   6   7   8   9   10   **AGREE**

_______ 1. I know how to identify project stakeholders.

_______ 2. I can determine stakeholder requirements.

_______ 3. I am good at identifying project deliverables.

_______ 4. I can prepare a project charter.

_______ 5. I understand some of the common reasons for project failure.

_______ 6. I know the critical factors to consider before starting a project.

_______ 7. I am confident in my ability to define the scope of a project.

_______ 8. I am familiar with PMI's Scope Management Model.

_______ 9. I understand what is meant by managing a project's critical first 10 percent.

_______ 10. I am good at developing a work breakdown structure (WBS).

_______ **Total**

Take a moment to reflect on the following before you move forward in the chapter.

What problems or situations have you experienced or observed related to defining a project scope?

Take a few minutes to reflect on your self-assessment. List two or three areas you want to improve.

As you progress through this chapter consider what actions you can take to demonstrate competence in the following three competency areas:

1. **Planning and Evaluation**. Establishes policies, guidelines, plans, and priorities; plans and coordinates with others; aligns required resources; monitors progress and evaluates outcomes; improves organizational efficiency and effectiveness.

2. **Resource Management**. Demonstrates awareness of technical resources; knows how to apply resources to achieve desired outcomes.

3. **Communication**. Makes clear and effective presentations to individuals and groups; listens to others; communicates effectively in writing; can critically review and comprehend information written by others.

Identify Project Stakeholders

After a project idea is accepted, the next steps are to identify the stakeholders, determine how to communicate with them, and then to define and formalize the project's scope. These efforts contribute to the overall goal of developing the project plan.

The illustration below shows these jobs are part of the Initiate phase of the project life cycle.

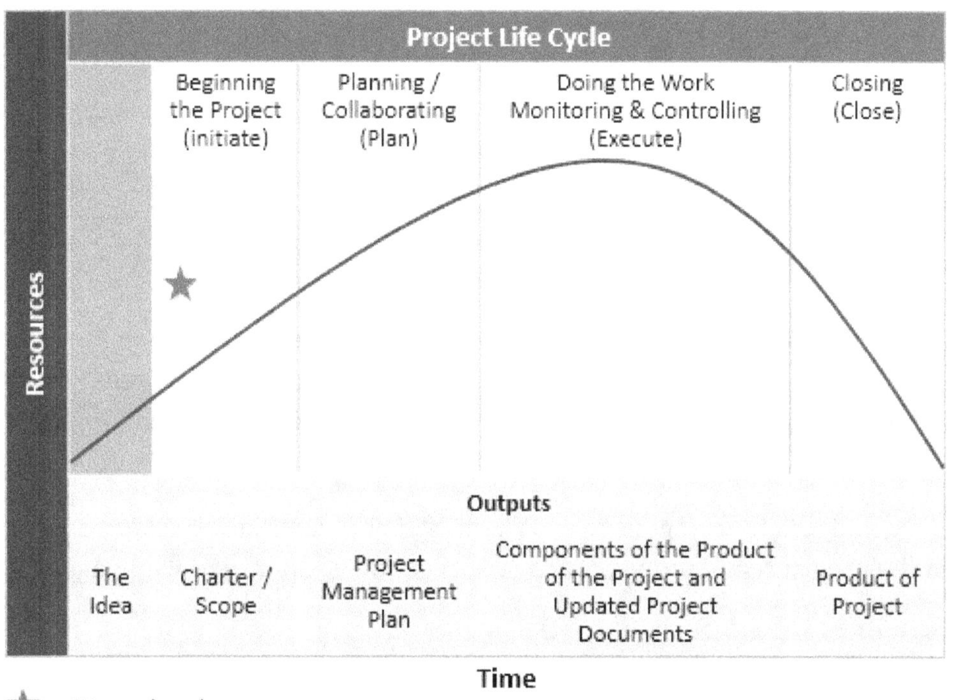

Also, be aware that changes in the project will occur throughout the project's life cycle. When they do, you must document them and revise the project scope and plan accordingly, and then communicate the changes to everyone involved.

Project Knowledge Areas and the Project Plan

The **Project Management Knowledge Areas** illustration below shows how the Project Management Plan enables integration among knowledge areas. The management plan is the mechanism that pulls all the knowledge area plans together.

Development of the Project Management Plan is accomplished by using what is called *progressive elaboration* or *rolling wave planning*. It means the Project Management Plan will require numerous iterations to stay current. As the project progresses, you must update the changes to the various knowledge area plans.

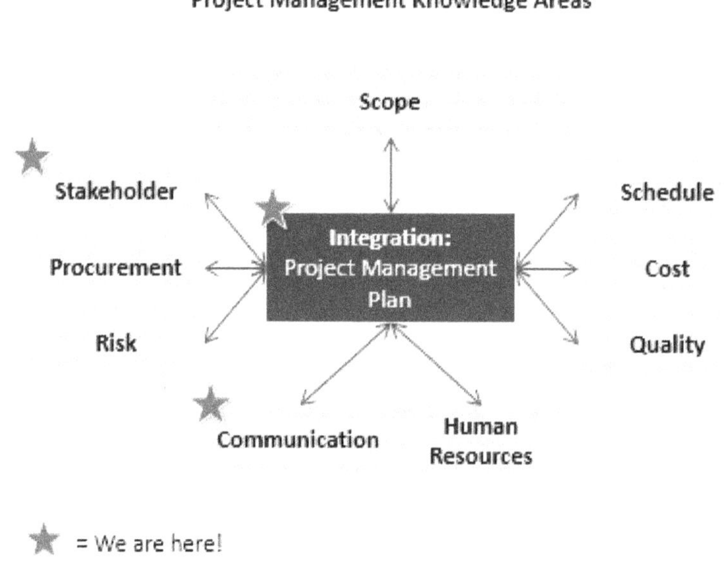

Identify Stakeholders Roles in the Project

A **stakeholder** can be anyone interested in a project's outcome. According to PMI, a *stakeholder* is "an individual, group, or organization that may affect, be affected by, or perceive itself to be affected by a decision, activity, or outcome of a project, program, or portfolio." (*PMBOK®Guide)- Sixth edition*. p. 723.

For example, in addition to the sponsor and client, stakeholders can be customers, members of the project team, project managers, executives, operational managers, end-users, and members of the public.

You must know who the project stakeholders are, what their roles will be, and determine how to communicate with them appropriately. Your goal is to communicate with all project stakeholders as early as possible, so their perspectives are included in the planning process.

PMBOK is a registered mark of the Project Management Institute, Inc.

Proper identification of stakeholders is a significant activity within the planning process. A Stakeholder Analysis Matrix is one way to identify the various stakeholders and their potential impact on the project.

Notice on the following stakeholder matrix example that the project manager has identified the impact each stakeholder may have on the project.

Stakeholder Analysis Matrix Example			
Stakeholder Name	**Stake/Interest**	**Potential Impact**	**Stakeholder Management Approach**
Stakeholder 1	Financial, ROI	Can approve or disapprove project funding	Meet early in the project regarding approval criteria. Likelihood of approval given specific metrics.
Stakeholder 2	Operations, processes may be impacted	Can provide supportive or negative voice	Include from the beginning to ensure this perspective is represented, and projected outcomes are positive.
Stakeholder *n*…			

Developing a Stakeholder Analysis Matrix and associated planning is sometimes an evolving process. For example, after a few iterations of the Project Management Plan, you might need to revise parts of it because a new stakeholder is identified. To update the plan, you would address any new tasks, risks, or resources.

To formally document the stakeholders for a project, use the above example as a guide. You will find a **Stakeholder Analysis Matrix Worksheet** in Appendix A.

To keep the analysis less complicated, sources of information for the stakeholder analysis are typically not identified in the matrix, but rather in a **Stakeholder Register.** A Stakeholder Register is also included at the end of this workbook in the Appendix Part B.

Example Stakeholder Register						
Stakeholder Name	Power / Interest Grid Position (Influence)	Org.	Org. Position	Location	Project Role	Contact Info.
Rebecca Smith	1	XYZ Rules	President	Washington, DC	Sponsor	rsmith@xyzrules.com
George Santos	4	Supplies Inc	Sales Rep.	Harrisburg, PA	N/A	sales@sales.com
Stakeholder n…						

Proper identification of stakeholders is a significant activity within the planning process. You want to be sure to include all potential stakeholders as early in the process as possible, so their perspectives are included in the planning process. It is also helpful to understand how much **power** and **interest** key stakeholder have related to a particular project as well as their desired level of engagement.

The intent of the Power/Interest, Power/Influence, or Influence/Impact **grids** are conceptually the same. You want to determine what kind of stakeholder(s) you are dealing with and how much attention you will pay to them. This analysis will contribute to the communication plan of how to communicate with each type of stakeholder, considering the content and frequency of the communications. If they are directly impacted by the product of the project, they should be in the regular communication circle of the project. If there is little change to their daily lives because of the outcome of the project, you would not want to bother them with frequent and detailed communications about the project. The following graphic is an illustration of a Power/Interest grid.

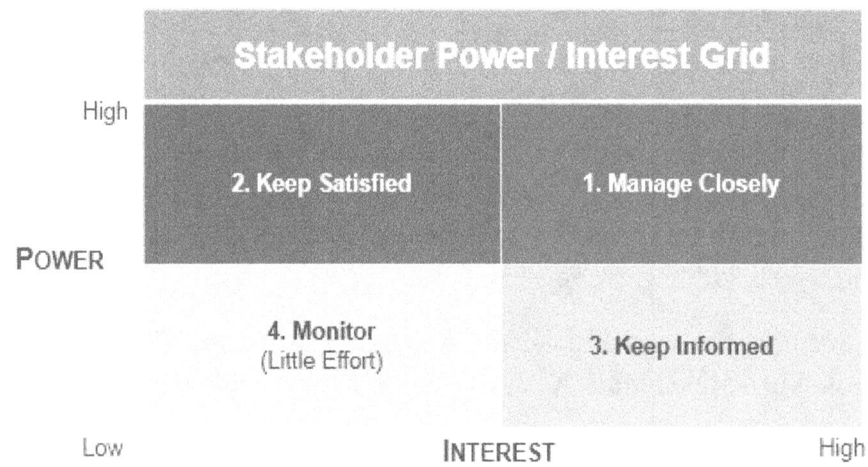

Another perspective of a stakeholder is their **level of engagement** in the project. It is important that key stakeholders are as **active** about the project as you had anticipated. There could be reasons for their lack of engagement that you need to know about as early as possible. First, determine what level of engagement you anticipate the stakeholder will have with the comings and goings of the project. As the project progresses, use the grid to assess the actual level of engagement as compared to the anticipated level of engagement. This periodic review can tell you that an important stakeholder has **drifted away** from the project, perhaps for negative reasons you should know about. The **Stakeholder Engagement Grid** provides a simple tool to keep this in check.

Example Stakeholder Engagement Grid					
Stakeholder	**Unaware**	**Against**	**Neutral**	**For**	**Promote**
Jim	C			P	
Jack		C	P		
Susan				P, C	
Ralph					P, C

P = Planned level of engagement (Plan Desired Actions)
C = Current level of engagement (Monitor)

A sample **Stakeholder Engagement Grid Worksheet** is included at the end of this course module as Appendix C.

Managing Communication

Communication can be one of the biggest challenges in any project. Managing communication from start to finish requires planning and discipline. The planning includes identifying the stakeholders, the types of information to be communicated, how the information is to be conveyed, and the frequency.

Communication is typically less complicated on small projects and so may not require as much planning and structure. But for larger projects, you need to plan who needs what information and how will you, as the project manager, deliver it to them.

Initial questions to consider might include: Who needs to know the project status? How will project status be communicated, for example, reports, meetings, or both? Who needs to know about project concerns or problems? Who can approve changes to the project? Does the project have a member directory? Is there a communication matrix for the project? If a stakeholder communication matrix is not available, it may be beneficial to develop one.

You must follow through with the communication plan throughout the life of a project.

What to do:	**Other actions:**
☐ Identify the stakeholders associated with your project.	☐ _______________________
☐ Create a Stakeholder Analysis Matrix.	☐ _______________________
☐ Assess the level of interest and engagement each stakeholder will have in the project and their potential impact on the project.	☐ _______________________ ☐ _______________________

- ✓ Stakeholders may be anyone interested in a project's outcome.
- ✓ Identifying and classifying stakeholders is a significant process to ensure that all stakeholders receive appropriate project communication.
- ✓ Stakeholders may change during the life of a project. When changes occur, you must update the stakeholder documentation.
- ✓ Communication is a critical factor in determining the probability of project success.

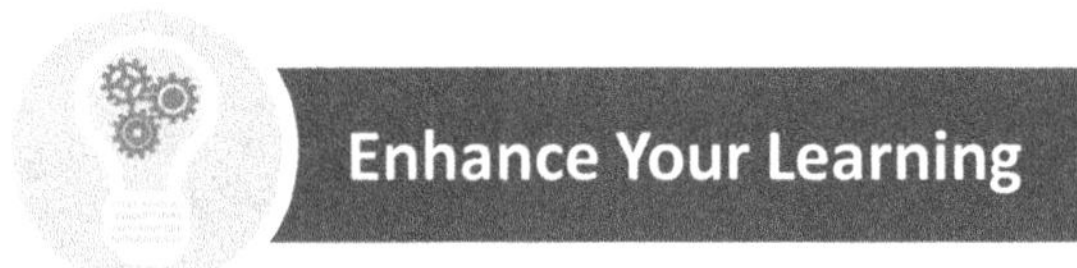

Watch the following 12-minute video by Arham Faraaz to learn more about identifying project stakeholders.

Faraaz, A. (2019). *Identify stakeholders | Project Stakeholder Management*.		Available at: https://www.youtube.com/watch?v=0xa2Q2XU2nM

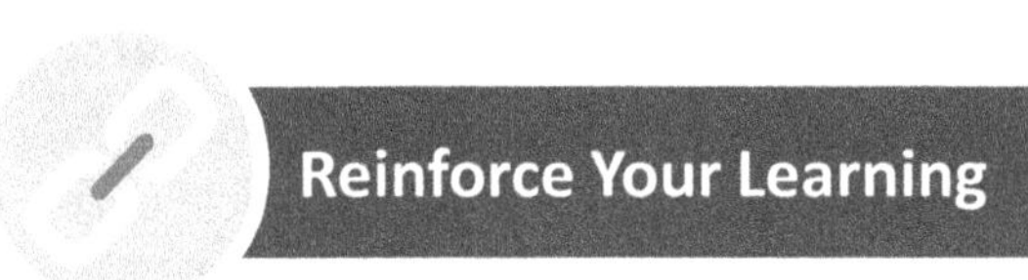

Stakeholder Analysis Matrix. Consider how the stakeholders in a project you worked on were engaged in the project and how their participation affected the success of the project. Then consider an upcoming project or one you are familiar with and complete the stakeholder matrix below. A Stakeholder Worksheet is included as Appendix A.

Stakeholder Analysis Matrix			
Stakeholder Name	**Stake/Interest**	**Potential Impact**	**Stakeholder Management Approach**
Stakeholder 1			
Stakeholder 2			
Stakeholder 3			
Stakeholder 4			

Determine Stakeholder Requirements

Defining and meeting the needs of the client and other stakeholders can be challenging. Various stakeholders may have requirements and priorities that differ one from another, and your job as a project manager is to help reconcile such differences.

Also understand that stakeholders can be *internal*, part of your organization, and *external*, outside your organization. You must meet the needs of both internal and external stakeholders. If you do not meet the needs of external stakeholders, the client may not return for repeat business, and your organization's reputation could suffer.

In a competitive bid situation, a potential client may issue a Request for Proposal (RFP). Anyone who has been involved in responding to competitive bids knows that RFPs can range from a few general ideas to specific and lengthy requirements.

After issuing an RFP, a potential client may hold a meeting for all bidders and other stakeholders, so people can ask questions. This is typically seen as the "fairest way" to deal with questions because everyone present hears the same information.

Whether an RFP is issued or not, a project manager must collect as much information as possible about stakeholder requirements. This strategy will help produce the details needed to define the project scope and the initial plan.

How to Determine Stakeholder Requirements

The first step in determining stakeholder requirements is to communicate and engage with them as early as possible. Do this as soon as you start work on the project. Start by identifying the key stakeholders, in other words, the stakeholders with the most power and influence, and document their requirements and what outcomes are most important to them.

Collecting stakeholder requirements is an essential part of completing a project charter and defining the project scope. The requirements need to be identified, recorded, and analyzed. Otherwise, you risk being unsuccessful because you will not understand what is to be accomplished, or you will misunderstand and attempt to provide something that is not wanted.

You can communicate with stakeholders and collect their requirements in a variety of ways. Here are some typical examples:

- Interview
- Focus groups
- Surveys and questionnaires
- Direct observation
- Pilot efforts and prototyping
- Historical information, data, and documents

You use the information gathered from these activities to identify and document stakeholder business requirements, constraints, assumptions, and expectations.

How Stakeholders May Judge a Project

The following is a summary of typical ways stakeholders judge how well a project meets their needs.

Quality
+ Technical quality
+ Timeliness
+ Effectiveness
+ Dependability/Reliability
+ Cooperation
+ Communication
+ Performance

Accuracy
+ Process
+ Information
+ Linking

Focus on Stakeholder Concerns
+ Their problems
+ Their convenience
+ Their perceptions

Promptness
+ Answering telephones
+ Information relay
+ Solutions
+ Keeping promises
+ Answers to questions
+ Progress reports

Attention to Convenience
+ Assuming responsibility for the outcome
+ Sensible, logical solutions
+ Reducing irritants
+ Efficient processes/solutions

What to do:

☐ Identify stakeholders and their requirements early in the project.
☐ Ask people who will do the work for their input.
☐ Provide opportunities for stakeholders to be engaged in the project.

Other actions:

☐ _________________________________

☐ _________________________________

☐ _________________________________

☐ _________________________________

- ✓ Consider the diverse needs of your stakeholders and be receptive to their feedback.
- ✓ Understand the concept of internal and external stakeholders.
- ✓ Stakeholder engagement includes attracting and involving individuals, groups, and organizations that may be affected by a project or may affect the project.
- ✓ Identify the key stakeholders and what outcomes are most important to them. Provide opportunities for them to be involved in defining the project scope.

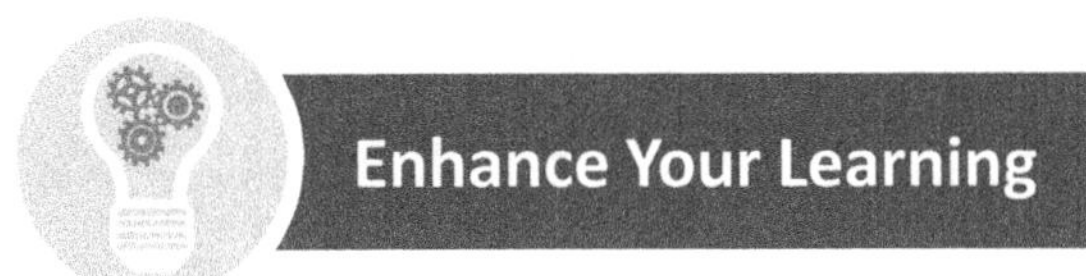

Review the following website by Harry Hall, The Project Risk Coach, to learn ways to engage stakeholders.

Hall, H. (2022). *10 Ways to Engage Project Stakeholders.*		Available at: https://projectriskcoach.com/engage-project-stakeholders/

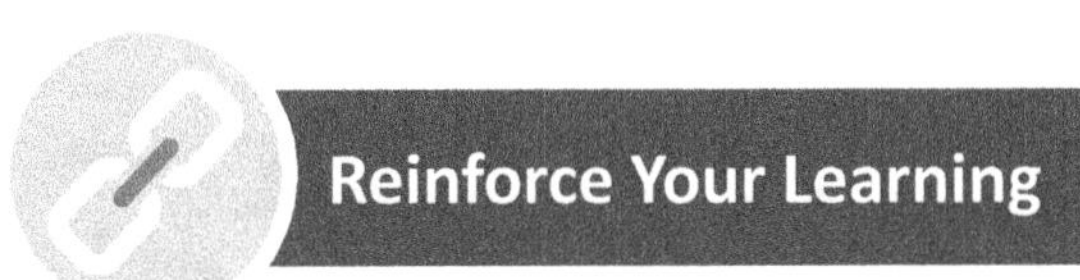

Stakeholder Engagement. Consider a project you were involved in or observed and recall how you worked with the stakeholders. Then answer the following questions:

Did you identify key stakeholders early in the project?

Did you provide opportunities for stakeholders to talk with each other and with you?

Did you listen to stakeholders and seek to understand their requirements?

Did you ask knowledgeable people for their input?

You want to be extra rigorous about making the best possible thing you can. Find everything that's wrong with it and fix it. Seek negative feedback, particularly from friends.

—Elon Musk

A PROJECT DELIVERABLE CAN BE ...

2.3

Specify Project Deliverables

Many inputs and outputs are associated with the completion of a project. Understandably, the most essential deliverable that stakeholders expect is the project's product, service, or result satisfactorily completed on time and within budget.

But to successfully plan, control, and deliver the project's product, service, or result, you must formalize project information, decisions, and changes by creating and communicating many documents. You must do this throughout the project's life cycle. You typically provide project documents to stakeholders and team members.

Early in a project, one of your first jobs as a project manager is to identify the types of documents the stakeholders and team need, how you will provide the information, for example, in paper or online reports, and the schedule, for example, weekly.

These document **deliverables** (both intermediate and final) serve as **communication tools**.

The following is an overview of the types of project documents you will use in most projects:

- **Project plan** or **work breakdown structure (WBS)**. It defines the project deliverables and the tasks/activities needed to complete each deliverable.
- **Weekly project meeting minutes**. The meeting minutes document formal project discussions and provide a record of issues/problems, actions, changes, and plans.
- **Project status, informed update,** or **variance report**. Use status reporting to aid the team in communicating progress, obstacles, and other pertinent information as the project progresses, and for you to communicate this information to stakeholders. Milestones are established and communicated in these reports.
- **Timesheet report**. This report documents each individual's hours worked on a project, and consolidated reporting shows time worked by selected project groups or functions.
- **Risk management plan**. This plan specifies project contingencies and how the team will avoid or cope with them. This plan helps the project manager and team minimize an organization's exposure to legal, financial, and public relations issues.

- **Project budget** or **financial report**. The organization uses this information to track and control project costs.

- **Communication plan**. Developing a plan for how to communicate project information among team members and among the team and stakeholders is an effective way to set up efficient channels for the exchange of project information, help maintain team morale, and support better client/stakeholder relations.

- **Lessons Learned.** After the team completes the project's product, service, or result that meets the quality level agreed on, this report describes the experience gained, so that future projects may benefit from the knowledge. The report describes both successes and failures. Additionally, the Project Management Institute (PMI) recommends that project managers capture lessons learned throughout the project life cycle. This encourages continuous improvement and leads to a more successful project.

- **Final report**. This report wraps up the project. It provides any final documentation to the people who need it, such as the team, client/stakeholders, and any other people in the organization or the public.

A stakeholder register can be an additional elaboration of a stakeholder analysis matrix that includes attributes for each stakeholder, including what deliverables are most important to each stakeholder or stakeholder group.

What to do:	**Other actions:**
☐ List the documents that will be used in most projects.	☐ _________________________
☐ Establish channels of communication for the exchange of project information.	☐ _________________________
☐ Communicate progress, obstacles, and other pertinent information on an ongoing basis as appropriate.	☐ _________________________
	☐ _________________________
	☐ _________________________

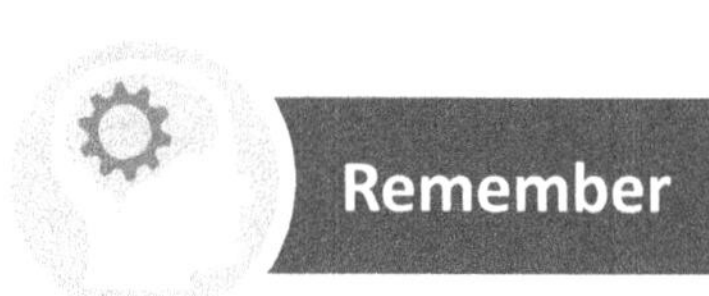

- ✓ Be proactive in providing information to stakeholders.
- ✓ Many inputs and outputs are associated with a project, and project managers typically provide these to team members and stakeholders.
- ✓ The most crucial deliverable that client/stakeholders expect at the end of the project is the product, service, or result satisfactorily completed on time and within budget.

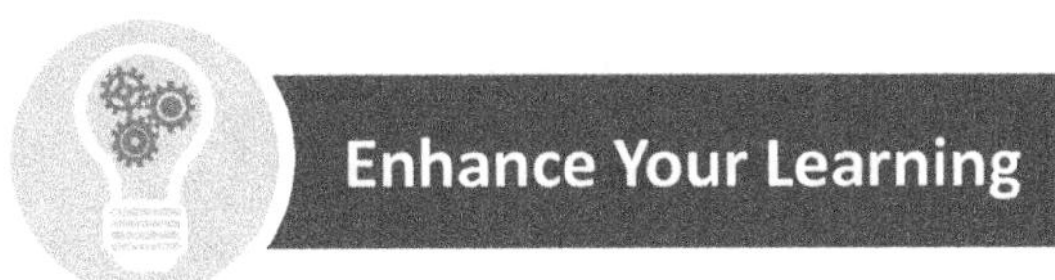

Enhance Your Learning

Watch the following 3-minute video by Chris Burke of PM Academy that describes how to identify project deliverables.

Burke, C. (2021). *How to Create Effective Project Deliverables.*		Available at: https://www.youtube.com/watch?v=7HADacRHSFs

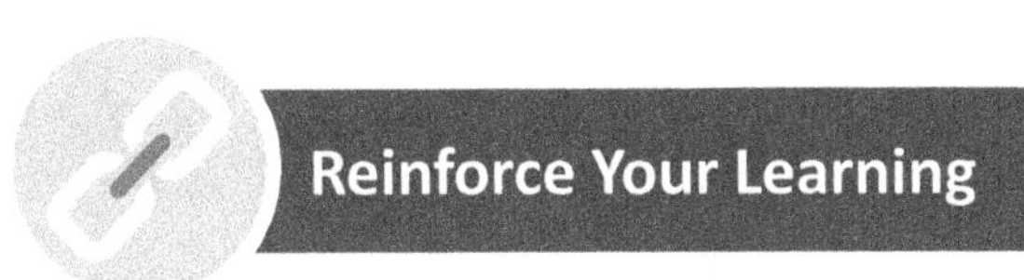

Reinforce Your Learning

Stakeholder/Project Deliverables and Communication Tools Matrix. Assume you must develop a stakeholder deliverables matrix like the sample below. For each stakeholder listed, place an "X" in the spaces provided to indicate the deliverables you believe the stakeholder may expect or benefit from

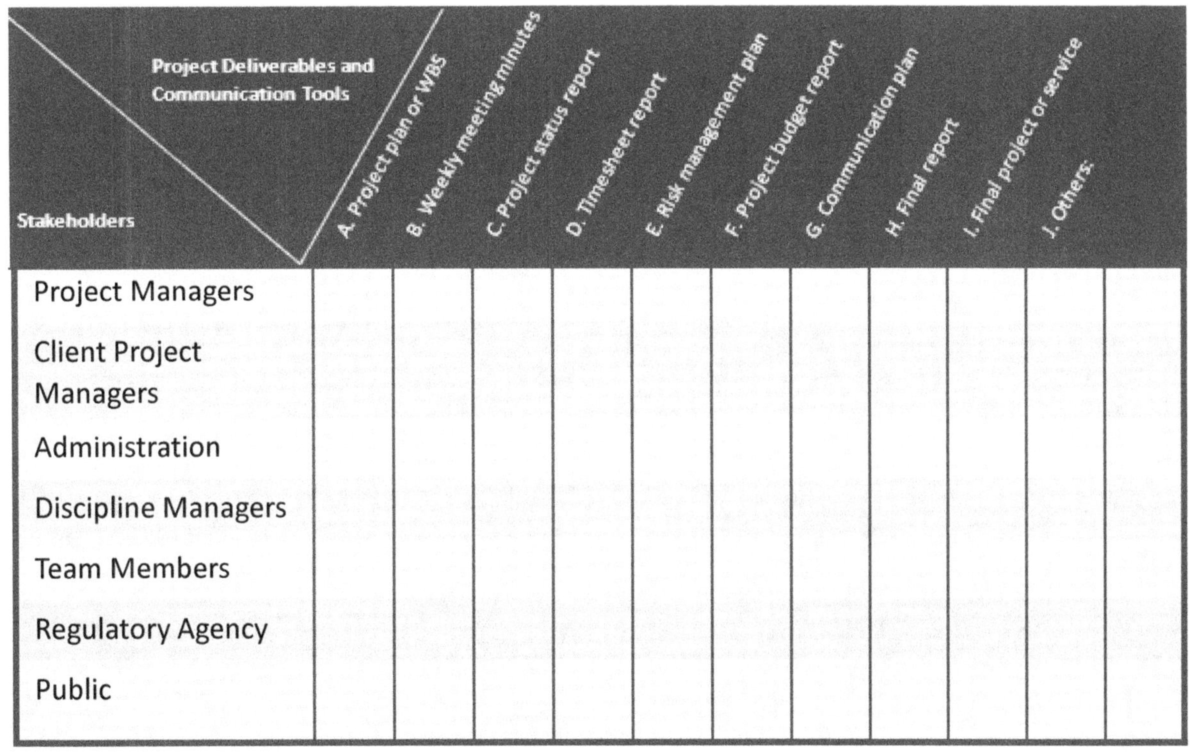

Stakeholders / Project Deliverables and Communication Tools	A. Project plan or WBS	B. Weekly meeting minutes	C. Project status report	D. Timesheet report	E. Risk management plan	F. Project budget report	G. Communication plan	H. Final report	I. Final project or service	J. Others.
Project Managers										
Client Project Managers										
Administration										
Discipline Managers										
Team Members										
Regulatory Agency										
Public										

Good business leaders create a vision, articulate the vision,
passionately own the vision, and relentlessly drive it to completion.

—Jack Welch

Draft Project Charter

According to *PMBOK® Guide*, A **project charter** is a document issued by the project initiator or sponsor that formally authorizes the existence of a project and provides the project manager with the authority to apply organizational resources to project activities. Often the **project charter** delineates roles and responsibilities, outlines major project deliverables, identifies the main stakeholders, and specifies the authority of the project manager.

A project charter is especially useful for larger projects. For smaller projects, a formal quote, purchase order, or Statement of Work (SOW) may suffice.

The project charter is typically a succinct formal high-level document that conveys the overarching purpose of a project and the expected product, service, or result. The document usually identifies the project manager and authorizes the expenditure of funds for the project.

The project sponsor usually authorizes or signs off on the project charter.

A project charter typically includes the following kinds of information:
- Project title
- Project manager
- Purpose or justification
- Objectives (address a need, improve a process, fix a problem, pursue an opportunity)
- Success criteria (improve process efficiency by X%, reduce costs annually by $ X, increase sales by X percent, reduce customer complaints by X percent):
- High-level project description
- High-level requirements
- Product characteristics
- Project constraints and assumptions
- Summary milestone schedule
- Summary budget
- Project approval requirements (success metrics, who decides, who signs off)
- Project manager authorizations (up to (x) $ signature; up to (y) $ change controls)
- Client or Sponsor (name, title, date)

As stated previously, the **project charter** formally recognizes the existence of the project. It gives the project manager the authority to proceed with the project, so it should be issued by a person or group with the authority to sponsor the project.

<table>
<tr><td>

What to do:

- ☐ Determine what level of authority you have to initiate projects in your organization.
- ☐ Discuss whether a quote, purchase order, or statement of work (SOW) is sufficient authorization to initiate projects.
- ☐ Identify who in your organization has the authority to sponsor or authorize large projects.

</td><td>

Other actions:

- ☐ ___________________________
- ☐ ___________________________
- ☐ ___________________________
- ☐ ___________________________
- ☐ ___________________________

</td></tr>
</table>

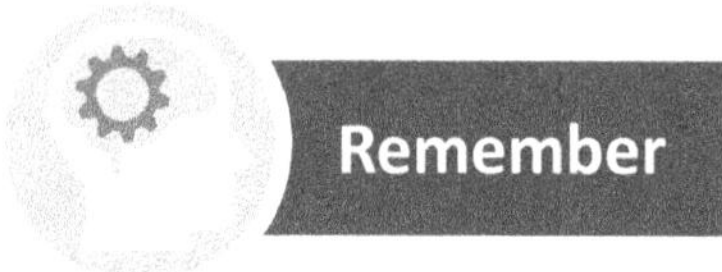

✓ A Project Charter formally recognizes the existence of the project and gives the project manager authority to begin the project.

✓ A project charter is issued by a person or group with the authority to sponsor the project.

✓ For smaller projects, a quote, purchase order, or statement of work may suffice.

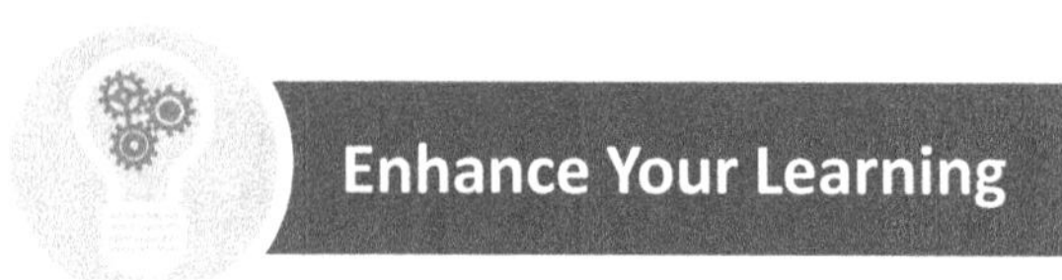

Watch the following 10-minute video by Adriana Girdler to learn more about preparing a project charter.

Girdler, A. (2020). *Project Charter Guide: HOW TO WRITE A PROJECT MANAGEMENT CHARTER.*	Available at: https://www.youtube.com/watch?v=FIPeKqrvNCs

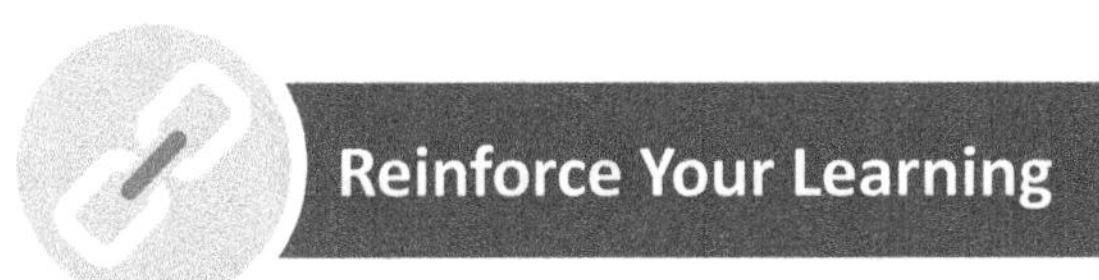

Project Charter. Examples of projects in the workplace include installing equipment, developing or conducting training, developing a new product, planning a significant event, and relocating an office or modifying building facilities. Select one of these examples or another that you are familiar with. Then complete a Project Charter for the project using the template below. If necessary, modify it to meet your needs. A Project Charter Worksheet is included at the end of this workbook as Appendix - Part D.

Project Title:

Project Manager:

Purpose or Justification:

Project Goals or Objectives (address a need, improve a process, fix a problem, pursue an opportunity):

Success Criteria (improve a process efficiency by X percent, reduce costs annually by $ X, increase sales by X percent, reduce customer complaints by X percent):

High-level Project Description and Requirements:

Project Constraints and Assumptions:

Summary Milestone Schedule:

Summary Budget:

Project Approval Requirements (success metrics, who decides, who signs off):

Project Manager Authorizations (up to (x) $ signature; up to (y) $ change controls):

Client or Sponsor: _____________________ Title: _________________ Date: _________

Those who plan do better than those who do not plan,
even though they rarely stick to their plan.

—Winston Churchill

Learn Common Causes of Project Failure

Are you aware of some of the classic project failures that have occurred? They include the Challenger Space Shuttle Disaster, New Coke, and even the Leaning Tower Pisa. Can you think of others?

Many lawsuits are filed against businesses and individuals because project issues were glossed over, or the planning was poor. Just as planning a project is critical to its success, so is revising the plan as changes occur over the life of the project.

In this concept, we will discuss reasons why projects fail. Projects are complicated endeavors, and especially large projects with extended schedules. An example is when a project team involves people who may not have worked together before, and they are doing something new.

Projects fail for many reasons, with inadequate planning and poor communication being among the most significant issues. The following are some other typical reasons:

- Business requirements not identified or communicated
- Project not adequately vetted or authorized
- Project expectations not aligned with organizational goals
- Priorities, objectives, and scope constantly changing
- Assumptions poorly defined and unsubstantiated
- Feasibility analysis not fully developed or inaccurate
- Tools and systems outdated and inaccessible
- Team collaboration inadequate; team conflicts
- Project leadership inexperienced and ineffective
- Resources untrained and misaligned

Also, recognize that people's actions cause the majority of project problems. It has been suggested that **people's** actions make up about **90 percent of the risks** in a typical project's Risk Management Plan (Bourne, 2009).

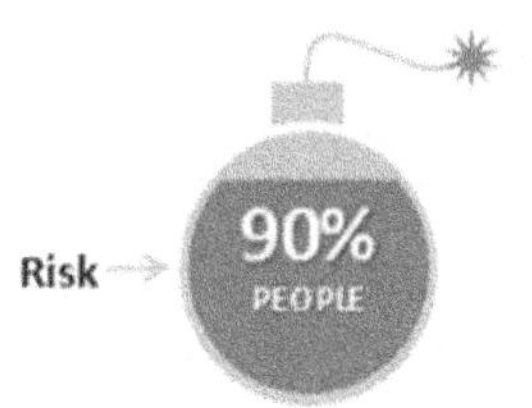

Among the biggest reasons projects fail are **changes in an organization's priorities** and **changes in project objectives** (*PMBOK® Guide*, 2013). However, when you consider the myriad of stakeholders with their individual perspectives and expectations, it is not surprising that inaccurate requirements gathering, not defining risks, and communication are not far behind.

But no matter how many requirements, or how many competing requirements among stakeholders, and no matter how many changes to the requirements occur as the project moves forward, the project manager is responsible for ensuring that all requirements are documented and approved by the stakeholders. The project manager is also responsible for ensuring that all requirements are accounted for in the original project plan and that the project scope and plan are revised when requirement changes occur.

Communication and Project Failure

The 2015 *Pulse of the Profession* from the Project Management Institute (PMI) reports that inadequate/poor **communication** is among the top five reasons why projects fail. However, one could argue that incomplete/inaccurate requirements and undefined risks are linked to inadequate/poor communication.

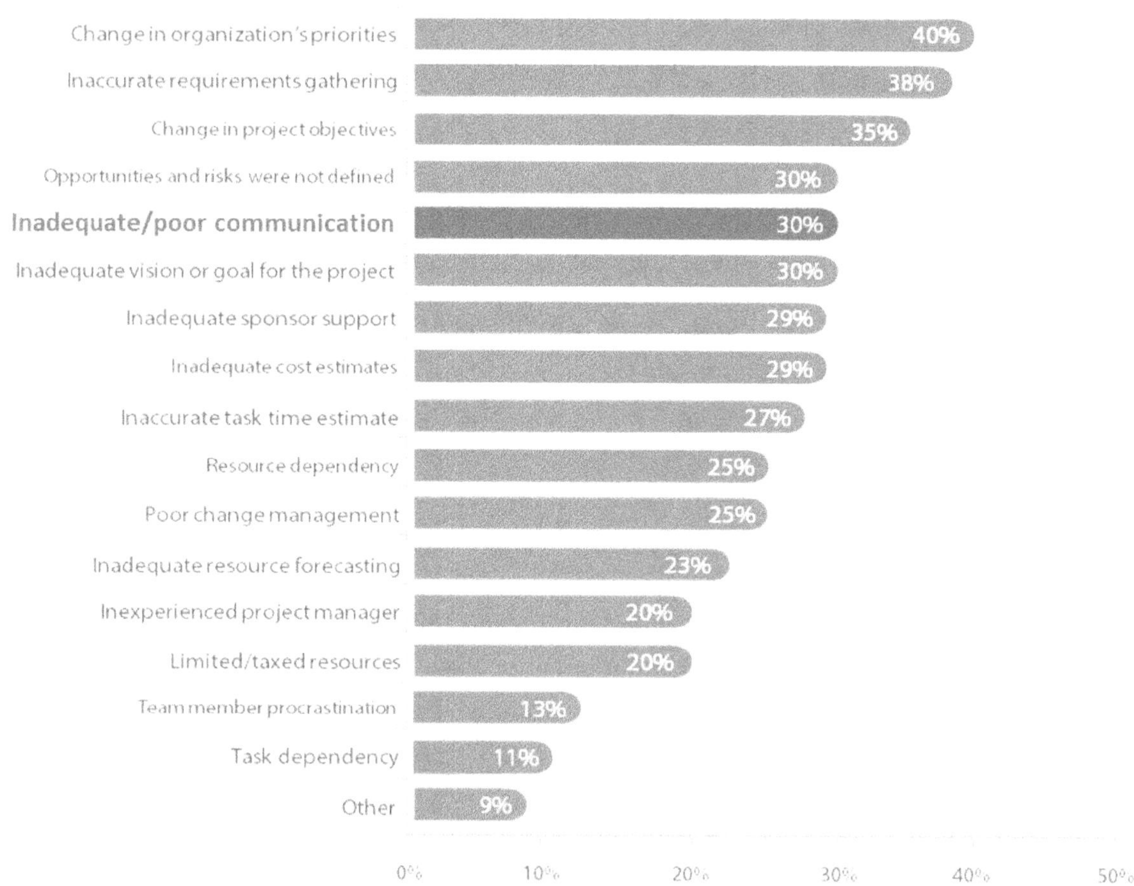

All these issues point to the importance of the initial project-planning phase and the need to accurately assess and define the scope of the project.

What to do:	**Other actions:**

☐ Recognize that people's actions may account for the majority of project problems.

☐ Consider if you are getting or have all project requirements.

☐ Consider the risks for your project and what could go wrong, and the impact on the project.

☐ ______________________________

☐ ______________________________

☐ ______________________________

☐ ______________________________

☐ ______________________________

✓ Among the biggest reasons projects fail are changes in an organization's priorities and changes in project objectives. This is followed by incomplete/inaccurate requirements gathering, not defining risks, and poor communication.

✓ People's actions make up about 90 percent of the risks accounted for in a typical project's Risk Management Plan.

✓ A project well done is always a project well begun.

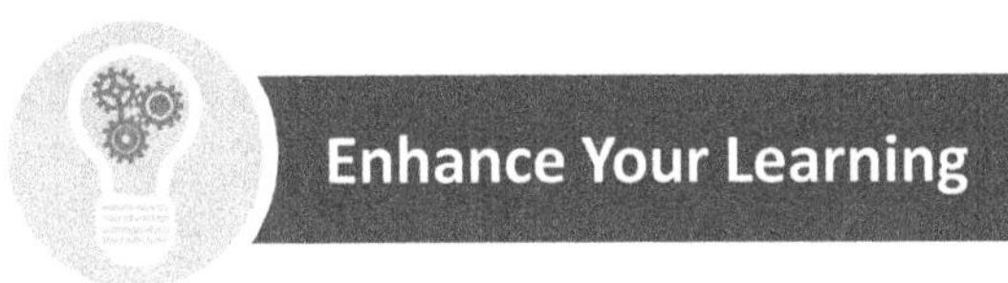

Watch the following 9-minute video by Jennifer Whitt to learn more about project failure. :

Whitt. J. (2012). *Why Projects Fail: Top 10 Reasons for Project Failure.*		Available at: https://www.youtube.com/watch?v=CTHHiBNXJ6w

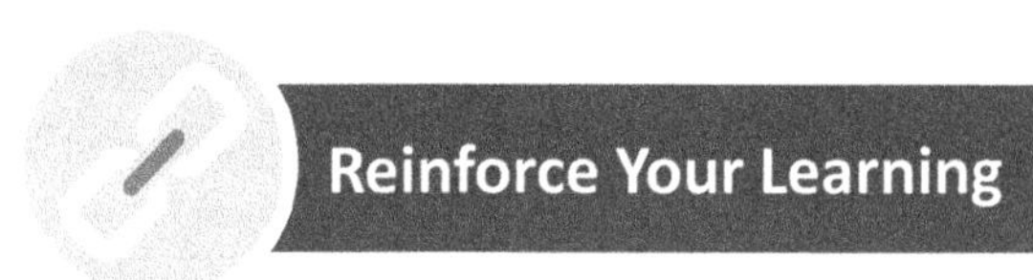

Why Projects Fail. Think of a project you have been involved with or observed that failed to achieve its goals. List three reasons for the poor performance and describe what the project manager could have been done differently.

1.

2.

3.

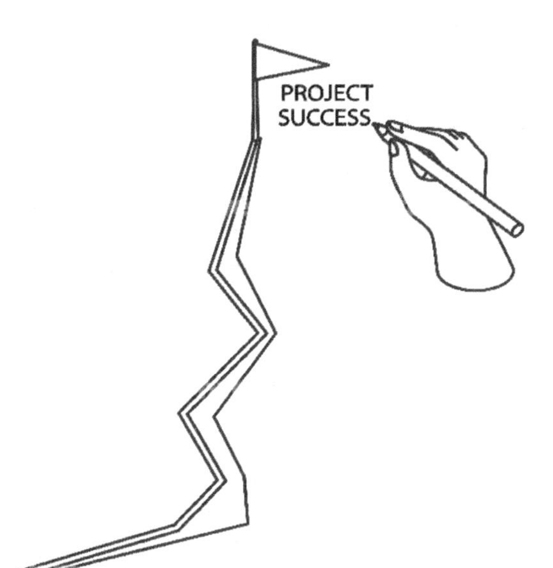

  **2.6**

Prepare for Project Success

In the previous concept, we discussed several reasons for project failure. In this concept, we will revisit the reasons and transform them into proactive actions project managers can take to help ensure project success.

You will benefit from these actions if you use them to prepare the groundwork for the project before you start work. These actions are the things you need to do in preparation for starting the project.

Many of these actions will also be productive at any time in the project life cycle when changes occur, and so you must revise or update project information. For example, if you learn of an additional business requirement while the project is in progress, you must go back to the stakeholders to define the requirement and get written approval from the client for the change.

Think of the information in this concept as a type of checklist. For example, the first activity discusses the necessity of gathering stakeholder requirements and why, but it does not describe how to do it. To learn more about *how* to do these activities, for example, how to gather requirements, refer to other concepts in this course.

The following is a discussion of 10 critical project activities that you as the project manager are responsible for completing.

1. **Identify and document business and client/stakeholder requirements.**

Be sure there is a strong business need for the project and that the stakeholders can support this need.

As a first step, gather the high-level requirements for the project, outline what you know already, and document what you think you are going to create. This will help you understand the background and context of the project.

You may need to gather requirements from a variety of sources, and it may take several meetings to achieve the final requirements. For example, for a software development project, two stakeholders, a call center manager and a marketing manager, give their high-level requirements. Then they may ask that the people gathering the requirements interview end-users in their organizations (subject matter experts) to get more details. You will need to ensure that these interviews are completed and that the requirements "hang together." If not, the people responsible for documenting the requirements must go back and review the situation with the operational managers to work out the differences and get stakeholder agreement on the final requirements.

If business needs or objectives change during the project, be sure to address this with the client/stakeholders so that you understand if and how the project plan needs to be revised.

2. Vet the project with the stakeholders and obtain proper authorization.
Work with stakeholders to outline a list of the project's expected benefits. With stakeholder buy-in and using the requirements information already developed in the first activity, project developers can then move forward with creating a high-level project description, goals, statement of work, anticipated outcomes, and justification.

You can then communicate this information to the project sponsor or group with the authority to approve the project. Be sure to get the approval in writing.

3. Align project expectations with organizational goals.

For larger projects, you will need to define the project's goals in a project charter. As we have discussed, the charter is a document that describes a project's purpose, and that gives a project manager authorization to start work. For smaller projects, you may need only a signed contract, statement of work, or purchase order.

Whether you use a project charter or one of the other methods, get the document in writing.

The project plan, as it evolves, should further explain the goals of the project and how they align, fit, or link to the organization's goals.

4. Manage project goals and objectives by clearly defining the initial scope and also define a change management process.

High-level goals and objectives are typically defined in the project charter. However, some will most likely change as the project progresses. The project manager and team refine the scope as the project evolves. The initial scope provides a benchmark as to what is and what is not included in the project from which the project manager can begin to show what the project's goal, deliverables, or finished product will look like at the end. As changes occur, they must be authorized, documented, and communicated using an established change process, so that everyone involved is kept informed.

Without such a process, scope creep may occur throughout the project, and this will lead to cost overruns, missed deadlines, and conflicts that are difficult to resolve.

A change management process lays out a framework for making changes, and the process must be communicated and understood by stakeholders because they are the ones who will approve the changes, along with changes in the budget and schedule.

5. Substantiate and verify project assumptions in the initial project scope statement.

Using the project requirements and any other information gained from the stakeholders, document project assumptions in the initial project scope statement. This information defines what the project must deliver and typically also provides information related to the cost, timing (milestones), and quality of the work products.

If no requirements were gathered, meet with key stakeholders to determine their requirements, so that assumptions, expectations, and the determinants of success can be verified, documented, and managed.

Again, scope creep is one of the most common reasons projects run over budget and deliver late. The client/stakeholders may forget the extra work and effort you have put in, insisting that you have achieved what they asked for originally.

6. Conduct a feasibility and risk analysis.

Many things can go wrong on a project. Some projects may benefit from pilot testing, but for others, that may not be practical. Taking the time to consider what your project may need is one of the best things you can do when you start a project.

Conducting an upfront and objective feasibility and risk analysis can help alleviate potential problems. Reference this information in the initial scope document and use it to highlight areas of concern and where prevention and mitigation strategies are needed. Involving key stakeholders and outlining what action you would take, should the worst happen, is of benefit to everyone involved.

7. Research and secure the proper tools and systems.

What are the right tools to use for a project? This depends on your project, your team, your client, key stakeholders, and your budget. Incorporating unnecessary tools may be costly and not give you the results you want.

Here are some commonly used tools:

- Resource planning and management: Float, Resource Guru, Hub Planner
- Project planning and timeline management: Microsoft Project, Gantt Pro, Workfront
- Collaboration with stakeholders: Google Sheets, Meeting Sphere, Confluence
- Communication with your team and key stakeholders: Zoom, Slack, Workspaces
- Managing project tasks: Trello, Jira, Monday.com

The best advice is to *keep things simple*. Consult with your team, client, and stakeholders. Avoid overcomplicating things as you can always integrate selected tools later if needed.

8. Provide adequate team building and development guidance.

If you have the luxury of selecting your team members, choose wisely. When you consider who might be available to work on a project, do not just look at availability. Think about what skills and experience levels are needed to deliver your project successfully.

Teams typically go through stages of development (form, storm, norm, and perform). Anyone who has worked in a group understands that building a high-performing team will take time and guidance.

9. Match project needs with effective project leadership.

Remember, 70 to 80 percent of employees quit their boss, not their job. A project manager must not only have some ability or technical skills in the project area, but must also be effective at managing, motivating, and leading team members. Leaders need to fit with the culture of the project team, client, and key stakeholders, and leaders should be selected that have experience completing projects of similar design, scope, and complexity without being overwhelmed.

10. Orient, align, and coach project workforce resources.

A project staffing plan defines who will be on a project team and how much of a time commitment each person will make. In developing this plan, the project manager typically negotiates with team members and their supervisors on how much time each team member can devote to the project. In other cases, a project manager may not have much say in who is assigned to a project.

Establish an onboarding and orientation plan to help new team members understand expectations and their roles and responsibilities. In some cases, individual skill assessments and specialized training may be needed. Communication is key. Ask for input, train as needed, provide guidance and coaching when needed, and tell your team when they are performing well.

What to do:

- ☐ Establish and agree on project boundaries before starting a project.
- ☐ Document project expectations for cost, scope, timing, and quality.
- ☐ Identify the actions you should take before starting a project.

Other actions:

☐ __________________________

☐ __________________________

☐ __________________________

☐ __________________________

☐ __________________________

Remember

- ✓ Identify and communicate business needs and requirements.
- ✓ Communicate with stakeholders and obtain proper authorization.
- ✓ Align project expectations with your organization's goals.
- ✓ Manage changes by defining the initial scope and set up a change management process.
- ✓ Substantiate and verify project assumptions in the initial project scope statement.
- ✓ Conduct a feasibility study and a risk analysis.
- ✓ Research and secure the appropriate methods, tools, and systems.
- ✓ Provide team-building and development guidance.
- ✓ Match project needs with effective project leadership.
- ✓ Orient, align, and coach project workforce resources.

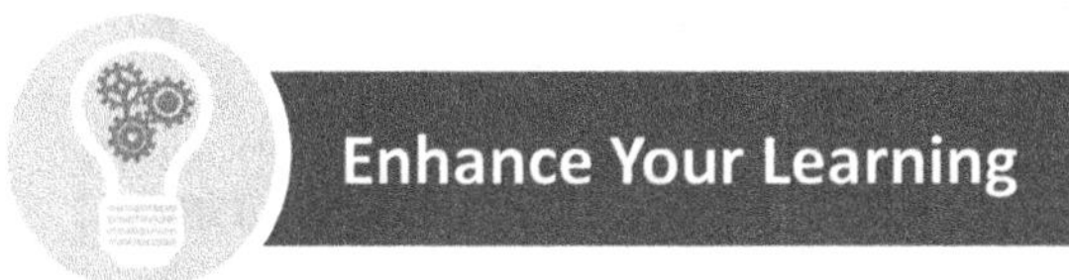

Enhance Your Learning

Review the article *"Top Things You Should Consider Before Starting a Project"* by Project Vanguards.

Project Vanguards. (2021). *Top Things You Should Consider before Starting a Project.*		Available at: https://projectvanguards.com/top-things-you-should-consider-before-starting-a-project/

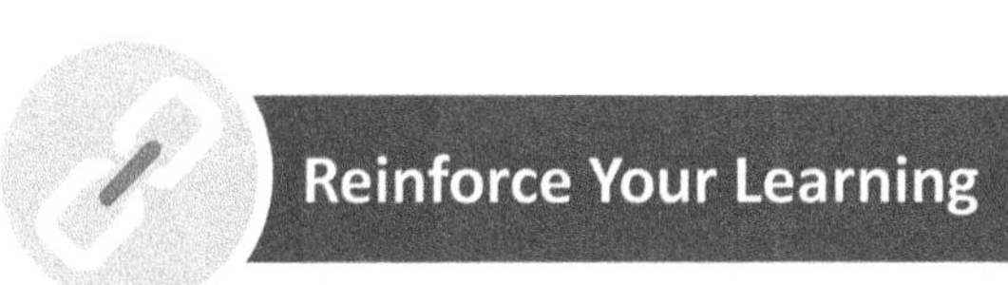

Reinforce Your Learning

Critical Factors to Consider Before Starting a Project. Consider a project you have been involved with or observed that failed to achieve its goals. List three things you would do differently at the beginning of the project to achieve better results.

1. ____

2. ____

3. ____

Of all the things I've done, the most vital is coordinating the talents of those who work for us and pointing them toward a certain goal.

—Walt Disney

Define Project Scope

When planning a project, the problem areas are always the same: costs, time, human resource failings, and the unknown. Thus, the more you know about the project before it begins, even before you develop action plans, the better.

According to *PMBOK®Guide*, a ***project scope statement*** is the description of the project scope, major deliverables, assumptions, and constraints. It is a document that describes what is in a project and what is not. The ***project scope*** is the work performed to deliver a product, service, or result with the specifies features and functions. A ***scope of work*** is typically a narrative description of the project.

The client/stakeholders typically delegate the creation of the initial scope statement document to the project manager. The project team assists the project manager in developing this document.

The initial scope statement document is often referred to as the ***scope baseline.*** Revisions are commonly referred to as a ***progressive elaboration*** of the scope. This term highlights that the characteristics of a project are determined incrementally and continually refined as the project progresses.

Project plan development takes the outputs from the project scope—and other planning processes—and combines them into a coherent and consistent document called the Project Management Plan. We will describe it in greater detail in later courses.

Prepare to Write a Project Scope Statement

Here are some critical questions you must answer before you start writing a project scope document:

- What is in the project?
- What is not in the project?
- How should the work be done?
- How much work should be done?
- What requirements do the client/stakeholders have for the project?

How the Project Scope Fits with Other Project Documents and Steps

The chart below illustrates the typical steps you will follow in creating the scope document. Notice that the Project Scope Statement is a deliverable in the Initiate phase.

Review each step listed under Description/Purpose and consider the reasons for each one.

Phase	Description/Purpose	Deliverables/Tools
Idea	• Preliminary description and analysis to determine whether a project idea has merit	• Idea appraisal • Decision: yes or no

Phase	Description/Purpose	Deliverables/Tools
Initiate: Define Project Scope	• Identify and document business need(s) • Define project scope, goals, and objectives • Identify project client, sponsor, stakeholders, and project organization • Develop initial estimates of effort, cost, schedule, and risk • Identify deliverables and acceptance criteria	• Project Description • Project Charter • Feasibility Analysis • Project Scope Statement • Stakeholder Analysis Matrix

Expect Changes to Occur as a Project Progresses

Recall that *project management* is a professional process that helps you manage organizational resources to achieve project goals within performance quality requirements, schedule, and budget or cost constraints.

As we have discussed, at the beginning of a project, you will define the project scope. But you can expect that some parts of the project will change as work progresses. If the changes affect the project scope, you must revise or refine the scope document to reflect the changes and communicate that information to the team and stakeholders.

But suppose you cannot deliver the expected or changed requirements within the established cost and scheduling constraints. In that case, as the project manager, you must meet with the appropriate stakeholders and determine how to modify expectations for the project and revise the project planning documents accordingly. Well-designed project scope statements can help you identify issues and problems and help you communicate them to stakeholders.

Audits of project results indicate that one of the most frequent reasons for project failure is an inadequate project scope document. This is because a poorly executed scope document does not describe what the client/stakeholders expect you to deliver. The project scope is the first significant step in defining and managing the project, so a poorly defined scope threatens the success of the project from the beginning.

As a project manager, it is your job to anticipate where a project might go wrong and to take corrective action when necessary. You cannot let your organization's investment slip through the cracks due to poor planning, mismanagement, or unexpected conditions. One of the best ways to stop this from happening is to develop a comprehensive project scope, keep it updated as the project progresses, and communicate the updated scope to the stakeholders and team.

Quality

The scope defines the final deliverable needed to satisfy project requirements. It is a rule of thumb that the client will want the deliverable—the product, service, or result—to be quality work. Therefore, the scope must also define the standards that define acceptable quality.

The client/stakeholders, project manager, and team members must understand and agree to the quality standards. Completing the work in a way that does not satisfy expectations is an almost guaranteed source of conflict and often leads to project failure.

Though they may not be specific, most clients/stakeholders tend to mean that they want high-quality when they say they want quality. Most project managers define *quality* as "meeting the requirements," though some managers may not take that a step further and make a distinction between "high quality" and "low quality." Take the time to understand and set up specific quality standards in the project scope document. It will help alleviate such miscommunication and head off misunderstandings and possibly failure.

Gold Plating

Adding extra features or components to a project beyond the original scope requirements is referred to as *gold plating.* While there can be a tendency for project managers to do extra things to please clients/stakeholders, be aware that adding to the requirements may affect the schedule and add to the project's cost. An example of *gold plating* would be if the initial requirements are for a Chevy and that is what the client is paying for, and then the project manager adds extras to give the client a Cadillac.

The initial, upfront effort a project manager takes to clearly define expectations, requirements, and standards in a project scope document will reduce the potential for conflict as the various project activities progress.

Communication

Communication is a concern and a priority in all project activities. As a project manager, you must communicate clearly. You cannot expect stakeholders to understand the jargon involved in the project. Project managers must take sufficient time and develop enough detail to fully complete scope documents that are understandable to all.

What to do:	**Other actions:**
☐ Meet with the stakeholders often to ensure you understand their requirements.	☐ __________________
	☐ __________________
☐ Communicate your understanding of the deliverables to the stakeholders.	☐ __________________
☐ Resolve any disagreement among stakeholders early in the planning phase and agree on the quality criteria for project success.	☐ __________________
	☐ __________________

Remember

✓ Project plan development takes the outputs from the project scope—and other planning processes—and combines them into a coherent and consistent document—the Project Management Plan. We will discuss it in later courses.

✓ Project scope planning involves creating the documents that define the project scope, and you may need to revise or refine it as the project progresses and changes occur.

✓ A narrative description of the work that must be done is called the Scope of Work.

✓ When planning a project, the problem areas are always the same: costs, time, human resource failings, and the unknown.

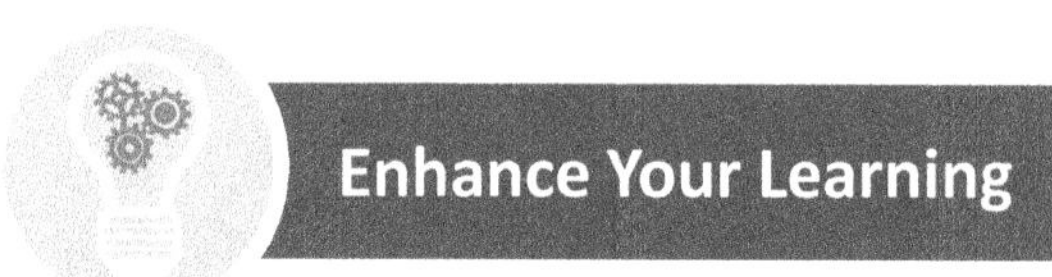

Enhance Your Learning

View the following 17-minute video by People Rich about scope management.

People Rich. (2012). *Project Management 2 – Scope Management.*		Available at: https://youtu.be/xjawa7wnd-w

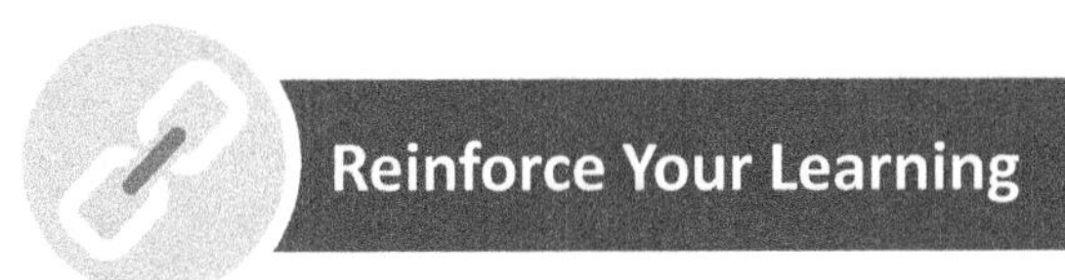

Items to Include in a Project Scope Statement. Consider the information you need to write a project scope document. Below are two project scope statements for constructing a house. Indicate the one you think is better and why. Then make suggestions about how to improve the project scope document.

Statement A: The construction project will include the following:

- Clear the land.
- Build the house.
- Landscape the area around the house.

Each of these steps should be done in a way that meets quality requirements.

Statement B: The contractor will:

- Clear the land to ensure that no tree stumps are visible above ground level.
- Build the house according to the attached design and material specifications the contractor and customer have agreed on.
- Landscape the entire property that surrounds the house once the building is complete. The landscaping will include two inches of # 2 sifted topsoil over the whole area and planting "select turf" grass seed.

Suggestions for improvement:

1. ____

2. ____

3. ____

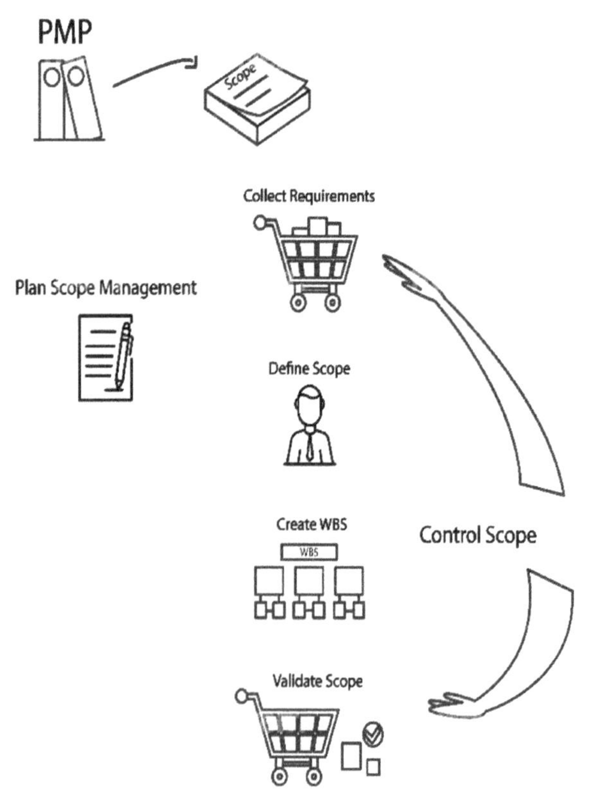

Review PMI's Scope Management Model

Project scope is at the heart of project management. As discussed previously, it is a description of the work that must be completed to achieve the project's goal, a new product, service, or result, and with the required features and functions, and according to the quality standards agreed to at the beginning of the project. The completion of a project is benchmarked against the project plan.

To avoid confusion, be aware that **project scope** and **product scope** do not mean the same thing. *Product scope* is a description of the features and functions of the new product, service, or result a project intends to deliver. The delivery of a product is benchmarked against the requirements for the new product, service, or result.

The **Project Management Institute (PMI)** offers a model for the scope management processes. It includes the following steps (Adapted from *PMBOK®Guide- 6th Edition*, p. 130):

1. Plan scope management

2. Collect requirements

3. Define scope: detail project deliverables as manageable components

4. Create a Work Breakdown Structure (WBS): subdivide the project deliverables into sub-deliverables and then tasks

5. Validate scope: get agreement among the stakeholders on project scope

6. Control scope: monitor and manage changes to the original project scope

The following chart lists the phases of the project life cycle that we have discussed, along with the corresponding steps listed above for PMI's scope management model.

Idea: Preliminary Project Description, Appraisal, and Decision to Pursue

Initiate: Begin the Project (1. Plan Scope, 2. Collect Requirements, 3. Define Scope; plan and develop project charter and initial or baseline scope statement document)

Plan: Plan and Collaborate with Key Stakeholders (4. Create a WBS; identify and detail project tasks or product components aligned with the project scope, develop a project management work plan)

Execute: Do the Work, Monitor, and Control, (5. Validate Scope, 6. Control Scope; verify the scope and update documents, manage execution efforts, manage scope change process)

Close: Close Project (Communicate completion of final project product or project, get agreement among key stakeholders)

In practice, many project activities and tasks overlap with interdependencies, and project life cycles vary, especially in Agile and Dynamic environments.

Agile project management environments, for example, software development projects, are characterized by frequent reassessment and adaptation of plans. In situations with emerging requirements, such as software development, less time may be spent on defining the initial baseline scope and more time in establishing an iterative process for scope refinement.

Related to Agile project management is Scrum, which focuses on short communication meetings, flexible reassessment of plans, and iterative phases of work.

What to do:	**Other actions:**
☐ Review PMI's Scope Management Model and associated concepts.	☐ ______________________
☐ Recognize that many project activities and tasks overlap with interdependencies and that project life cycles can vary.	☐ ______________________ ☐ ______________________
☐ Investigate how Agile and SCRUM techniques may help you manage a project.	☐ ______________________ ☐ ______________________

✓   The project manager is responsible for the project scope document.

✓   The scope document is the foundation of all other components of a project management plan.

✓   The initial project management scope document is often referred to as the baseline project scope.

✓   Agile project management environments are characterized by frequent reassessment and adaptation of plans.

✓   Related to Agile project management is Scrum, which focuses on short communication meetings, flexible reassessment of plans, and iterative phases of work.

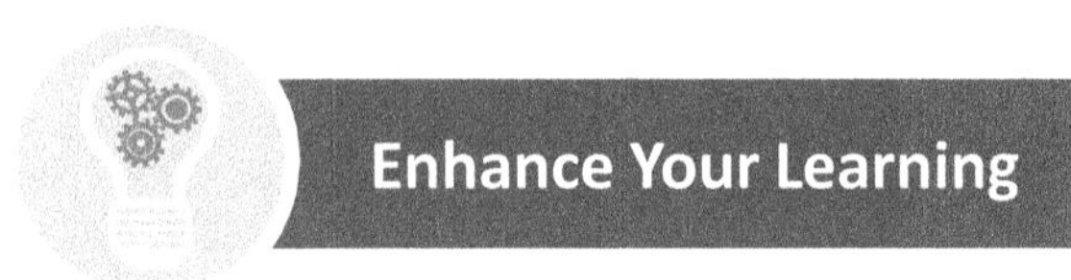

Watch the following 3-minute video from The Crowd Training to learn more about the Project Scope Management knowledge areas.

The Crowd Training. (2016). *Drawn Out Project Scope Management.*.		Available at: https://www.youtube.com/watch?v=KZn 2ly_p8pY

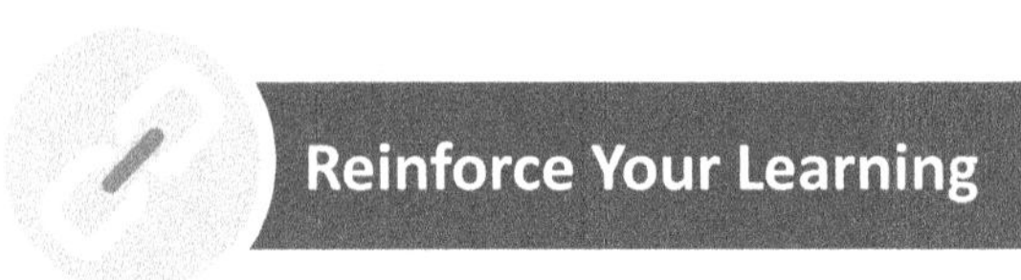

Project Scope Statement Document. The initial scope document is called the *scope baseline*. You will need to revise the scope baseline throughout the life of the project as changes occur. The revisions are referred to as *progressive elaboration*, which points to the fact that the characteristics of the project are determined incrementally and continually refined. Review the scope document template that follows and compare it to others you may be familiar with. Use this template to define the scope of an upcoming project or practice by focusing on a project that you are already familiar with. Appendix – Part E also has a worksheet you can copy and use for other projects.

Project Scope Statement Document

Project Title:

____

Introduction (overview of concept, schedule, and cost, and refined project objectives)

____

Description (overview of what is included and not included, and the cost-benefit analysis)

____

Deliverables (scope/schedule of work, deliverables, boundaries of what is and is not included, risks, and assumptions)

____

Resources (personnel and other resource requirements

____

____

Costs (all estimated costs and cost-benefit details)

____

____

Schedule (all estimated tasks/activities; Milestone chart)

____

____

Acceptance Criteria (success metrics for scope, costs, and schedule)

____

____

Client / Sponsor: **Title:**   **Date:**

____

Everyone has a plan until they get punched in the face.

—Mike Tyson

Understand Why Pre-planning is Critical

Project management professionals refer to the *critical first 10 percent* as the time spent planning a project from start to finish. This time is vital because everything that happens later in the project will be influenced by how well the project was planned.

The following is a summary discussion of the many responsibilities involved in planning a project. Review the summary and consider if you are prepared to meet these expectations.

Understand the demands of project planning. The work of a project manager requires you to think through every detail. You will need to ask many probing questions, starting with the who, what, where, when, and why of project planning:

- What benefits will the organization gain from the project?

- Who is the sponsor? The client? Other stakeholders?

- What product, service, or result do the client/stakeholders want?

- What are the stakeholder requirements for the project?

- What work needs to be done, in detail, from the start to the end of the project?

- Who are the people who have the skills and experience necessary to complete the project and will they be available in the time frames needed?

- What is the budget?

- Is there a deadline for when the project must be done?

- When is the project expected to start, and when will it finish?

- Why do you think you are qualified to lead the project?

As a project manager, you are responsible for ensuring that the time, money, and human resources of your organization are used wisely. Poor planning at the beginning of a project, regardless of its size, may lead to project failure and loss of time, money, and reputation.

Be aware that many planning documents are needed. As a project manager, you will develop several planning documents. You will also need to revise/update many of these documents as the project progresses and changes occur. The documents (both initial and final) typically include the following:

- Scope statement of the requirements for the project's product, service, or result, along with a description of the quality standards agreed to.

- Work breakdown structure (WBS) that defines the project deliverables and the activities/tasks needed to complete the project.

- Risk management plan that outlines all contingencies for the organization and how to avoid them or cope with them to minimize the organization's exposure to legal, financial, and/or public relations losses.

- Project budget that is an honest reflection of the demands of the project.

- Status report format and procedures the team can use to communicate progress on milestones, problems, and other pertinent information.

- Communication plan that supports team morale, client relations, and provides effective channels for the exchange of project information among all interested parties.

- Final report to the client/stakeholder.

Determine the project scope. As the project manager, you must envision the beginning and the end of a project and everything that happens in between. You must anticipate where things might go wrong and where time or money could be wasted. Thus, during the project planning stage, it is your responsibility to determine the scope and demands of the project from the creation of the contract to the collection of the last payment.

Chart the project's course: The project plan is the project's roadmap. Project managers must be efficient and effective. This is achieved by creating and executing according to a *project plan*, which is a realistic, action-oriented document with the following characteristics:

- Complete and current for the actual project, not a template plan

- Flexible and can be adapted to extenuating and unpredictable circumstances

- Logical and reflects the scope of the work to be done

- Not easy to ignore and will assist everyone in understanding the project and thus will be used

Break down work activities and tasks. Are you aware of the various project planning tools available? Examples of some useful tools include:

- Work Breakdown Structures

- Gantt Charts

- Network Diagrams

Be aware that a lack of planning will lead to poor results. Without a plan, you will not know if a project is going well or if it is in trouble. Further, even in a well-planned project, things will go wrong. As a project manager, you must expect and plan for glitches along the way. The key is to plan the project carefully and realistically, which will help you avoid as many problems as possible. When glitches happen, manage them so they do not grow into issues that can damage the project.

Budget and Control project costs. Your job as a project manager is to budget thoroughly and carefully, and then to monitor the budget as work progresses. You must estimate expenses and create a spending forecast. Think of it as a cash flow road map. You will not know where you are or where you are going without it.

As a project manager, you are the first line of defense in your organization's daily battle with marketplace competitors. To succeed, you need to ensure profitability. To do that, you need to know how much money is required to meet expectations and complete the project on time.

Start with a comprehensive analysis of the project and keep asking these questions: How much will be spent; where will it go, and where will it come from?

Avoid legal issues. Careful project planning is the best defense against litigation. Project managers must keep project objectives in mind, and this includes consideration of possible legal problems. It is often possible to avoid or minimize your organization's exposure to costly lawsuits by adding the proper legal clauses in project documents during the planning stage. Here, as in many areas, knowledge is power.

Are you aware of the many legal clauses to consider as you plan a project? Do you know how to incorporate them when appropriate?

Be prepared: No project is risk free! You plan for risk by developing a realistic assessment of potential problem areas and responses.

In a risk assessment, you answer two questions: Can your project team complete the deliverables, as promised, and what might get in the way of you presenting these deliverables to the client? Factors to consider include—but are not limited to—contract problems, vendor and/or subcontractor nonperformance, and the unavailability of resources.

Evaluate critical factors before starting a project. As a project manager, you are asked to make judgment calls throughout the proposal process. Your first judgment call is a "gut" reaction to a project opportunity or a Request for Proposal (RFP). You decide if it is worth investing the effort to progress to a pre-proposal face-to-face session with the potential client and a formal Go, No-Go decision process.

What to do:

- ☐ Be aware of what can go wrong as a result of poor project planning.
- ☐ Ensure that the project planning process meets client/stakeholder expectations and requirements.
- ☐ Recognize that careful project planning is the best defense against litigation.

Other actions:

- ☐ ___________________________
- ☐ ___________________________
- ☐ ___________________________
- ☐ ___________________________
- ☐ ___________________________

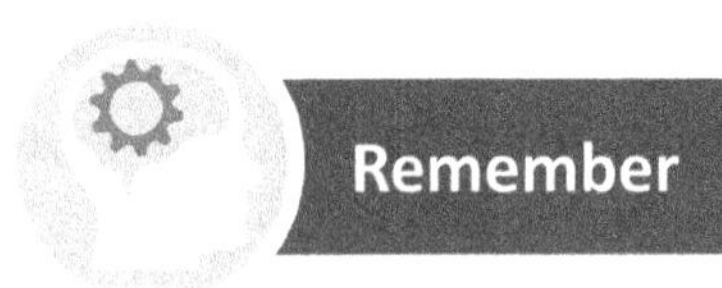

Remember

- ✓ When planning a project, the problem areas are always the same: costs, time, human resource failings, and the unknown. Thus, the more you know about the project before it begins, even before you develop action plans, the better.
- ✓ Every project, regardless of its size, scope, and budget, will face its share of setbacks. The key is to avoid problems that can wreck reputations, client relationships, the project, and even the organization itself.
- ✓ As the project manager, you must ensure that your project has a budget to cover all costs.
- ✓ The budget is one of the key considerations in project success.
- ✓ Often it is possible to avoid or minimize your organization's exposure to lawsuits by including proper legal clauses in documents created during the planning phase.
- ✓ The project plan is a guide. Involving others in the planning leads to more buy-in later in the project.

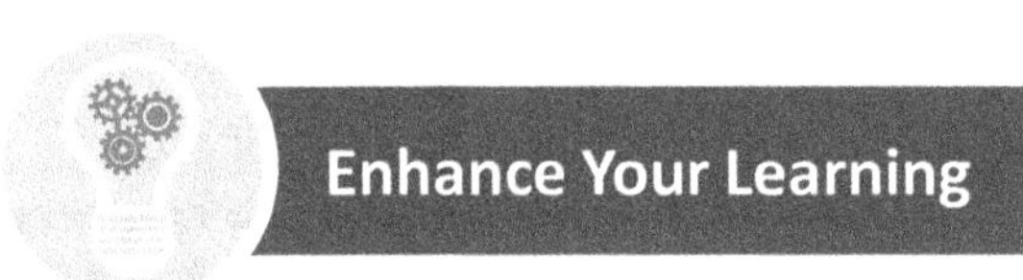

Enhance Your Learning

Watch the following 5-minute video by Jennifer Whitt to learn more about the project planning process. your work:

<table>
<tr>
<td>Whitt, J. (2013). Project Planning Process: 5 Steps to Project Management Planning.</td>
<td></td>
<td>Available at:
https://www.youtube.com/watch?v=Do8iykQKMfU</td>
</tr>
</table>

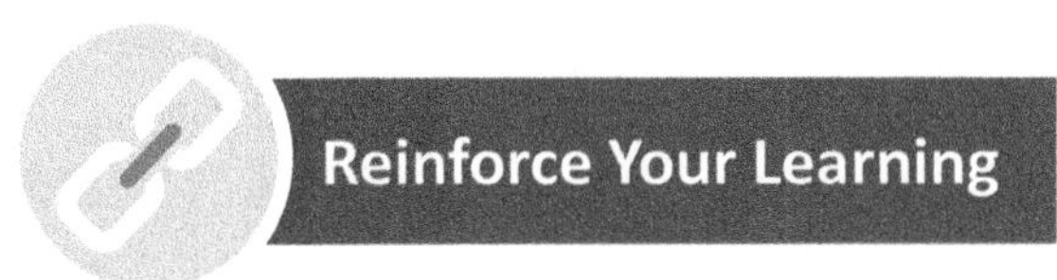

Critical Factors to Consider When Planning and Implementing Projects. Consider a project you are familiar with. Assume you are evaluating the critical factors discussed below to assess whether you can successfully complete the project. Review each of the statements and check the column that indicates your level of agreement.

How confident am I?	very little	some what	very much
I have planned and managed projects like this in terms of size, scope, and difficulty.			
I am comfortable that the scope of this project has been clearly defined and communicated to the team and stakeholders.			
I believe the budget established for this project is adequate and reflects reality. I can complete the project at or under its cap.			
The budget accurately depicts costs for the specific people who will work on the project.			
The proper people are on the project team. They have the right skills to meet the demands of the project.			
The team will stay together for the duration of the project.			
The project plan has clearly defined milestones, target dates, and deadlines. Ample time was built into the plan to allow for satisfactory completion of the project, meaning it will be on time and adhere to the budget.			

124

Even if you are on the right track, you will get run over if you just sit there.

—Will Rogers

Work Breakdown Structure

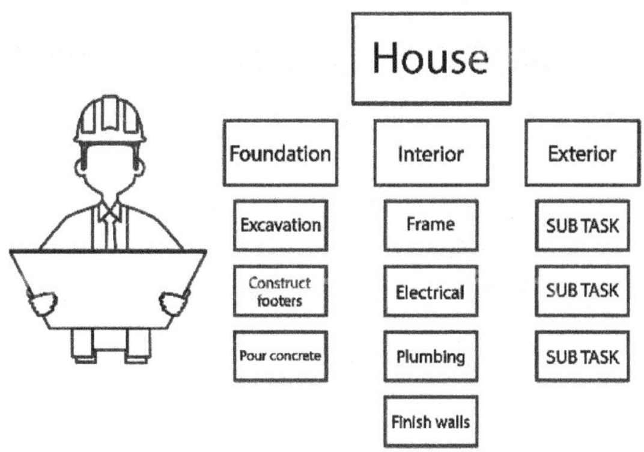

Develop Work Breakdown Structure

After the scope statement document, the next step in a project is to identify and define the work needed to achieve the project's product, service, or result.

You define the project's work by preparing a **Work Breakdown Structure** (WBS).

A WBS is a vital tool for project managers. In some cases, the WBS can be as simple as listing the tasks necessary to complete a project.

For more complex projects, a WBS is a detailed hierarchy of deliverables and sub-deliverables and the tasks needed to accomplish the deliverables.

A WBS structure provides management with a database for planning, executing, monitoring, and controlling the project work. The hierarchical structure also furnishes information appropriate to each level.

The following illustration is an example of a hierarchical WBS for planning a wedding:

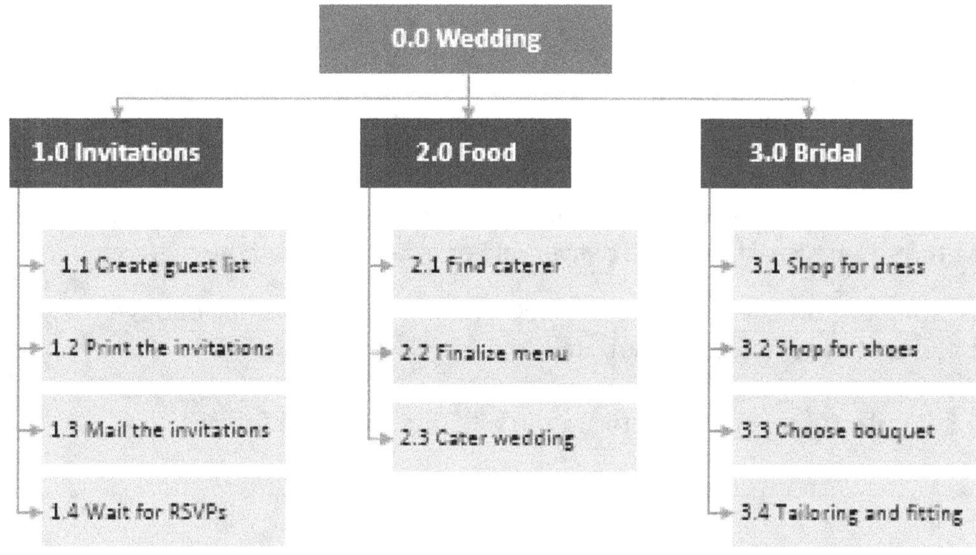

Start by Brainstorming Ideas for the WBS

Ideas for what is needed in a project's WBS are often generated in a **brainstorming** session facilitated by the project manager. It involves all the relevant people on the project team. Include those who know what needs to be accomplished and those who will do the work, and someone to take minutes.

The following are guidelines for running a productive brainstorming session:

- Ideas must be recorded publicly and as offered with no editing or paraphrasing.

- No one can comment (positive or negative) about the ideas offered. If needed, the minute taker may ask to have the idea repeated to ensure they record it correctly.

- The objective of a brainstorming session is idea quantity. After idea generation, the group evaluates, clarifies, and classifies the ideas to consolidate the list and reach some agreement on the best ideas.

- When necessary, the group can research information.

How to Develop a WBS

Work from the top down. Identify the deliverable and put the name in the box at the top. If it is a WBS for an entire project, put the project name in the box. In the example above, the deliverable is Wedding.

Then define the significant deliverables to achieve the project deliverable. In the Wedding example, these are Invitations, Food, and Bridal.

Note: If you want to make a WBS for a deliverable in the project, put the name of that deliverable in the box. You will likely also want to add its number. For more information about numbering, see the numbering section below.

Break each major deliverable into sub-deliverables. After you define the major deliverables, you will then identify the sub-deliverables necessary to accomplish each one. The total number of sub-deliverables depends on the complexity of the higher-level task.

Then break the sub-deliverables into tasks. Tasks are the lowest work level on a WBS. They represent work that cannot or does not need to be broken down further. A task must complete the sub-deliverable or the deliverable one step above in the hierarchy.

Keep repeating this process until the sub-deliverable details are small enough to manage, and one person can complete the sub-deliverable.

This hierarchical structure is illustrated in the WBS example that follows.

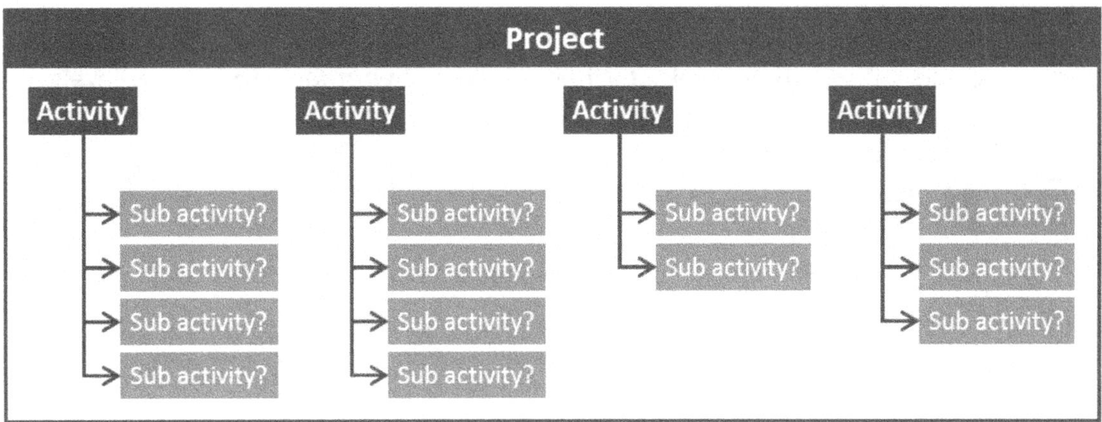

Work specifications can be written for each activity and sub-activity in the project.

You can also number each level in a WBS. See the WBS illustration below in the section on numbering.

If needed, you can develop a second WBS for some of the more complex sub-deliverables.

Work Packages

Sub-deliverables are then organized by **work packages**. These are typically the **lowest level breakdown** for a project manager. The work packages are grouped by the type of work involved, such as hardware, programming, testing, and so on.

The groupings are **cost accounts**. For example, hardware might be a cost account. This way of grouping by type of work supports a system of monitoring project progress by work and responsibility.

Here is a general **rule of thumb** for the size of a work package: **no work packet more than 80 hours of work effort or less than 8 hours of effort**.

The project manager defines the following information for each work package:

- Description of work (what)
- Time to complete the work package (how long)
- A time-phased budget to complete the work package (cost)
- Resources needed to complete a work package (how much)
- A single person responsible for the work (who)
- Monitoring points for measuring progress (control)

Number the Levels on a WBS

For larger projects with multiple levels, it is helpful to assign numbers to the sub-levels. Numbering is particularly useful when you use software tools.

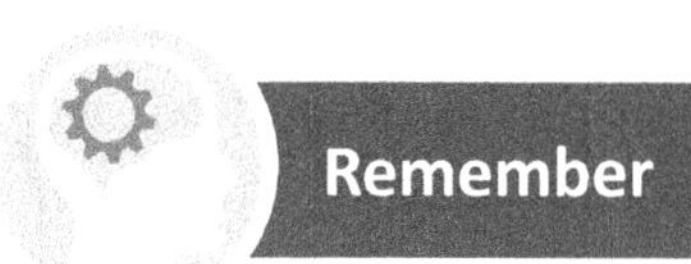

WBS Level 1 (1.0): Major project deliverables that are about same level of effort; equal 100%.

WBS Level 2 (1.1): Typically the primary sub-deliverables at which reporting is done.

WBS Level 3 (1.1.1): Work package level where cost and schedule can be estimated; 8hr. – 80 hr. rule.

WBS Level 4 (1.1.1.1): Activities that have to be done; where; who; when; how much $$$.

What to do:

- ☐ Identify project activities and alternatives by conducting a brainstorming session.
- ☐ Prepare a WBS that defines all the deliverables and work in the project.
- ☐ Break each deliverable into sub-deliverables. The number of sub-deliverables depends on the complexity of the higher-level task.

Other actions:

- ☐ ___________________________
- ☐ ___________________________
- ☐ ___________________________
- ☐ ___________________________
- ☐ ___________________________

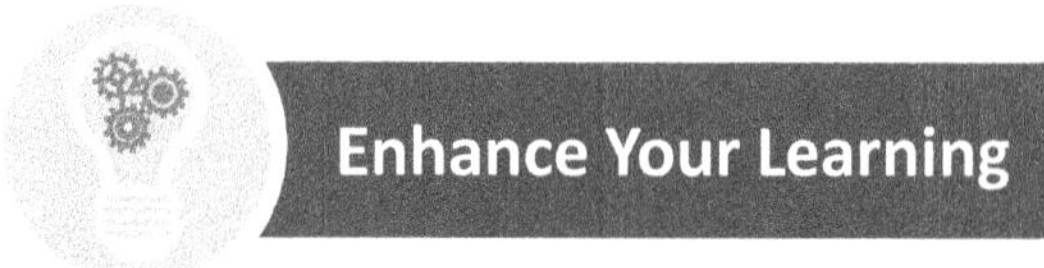

- ✓ Brainstorming is a useful tool for identifying project activities, generating alternatives, and obtaining input from stakeholders and team members who will do the work.
- ✓ Set a goal to meet or exceed stakeholder expectations and requirements.
- ✓ Prepare a WBS to include all the deliverables and work in a project. Then break each deliverable into sub-deliverables. The number of sub-deliverables depends on the complexity of the higher-level task. Repeat until the sub-deliverable details are small enough to manage, and one person can complete the sub-deliverable.

Enhance Your Learning

Watch the following 8-minute video by Dartmouth University about the development of a multi-level work breakdown structure.

Dartmouth University. (2014). *Five-Level Work Breakdown Structure.*		Available at: https://www.youtube.com/watch?v=SqcuLsyFr-o

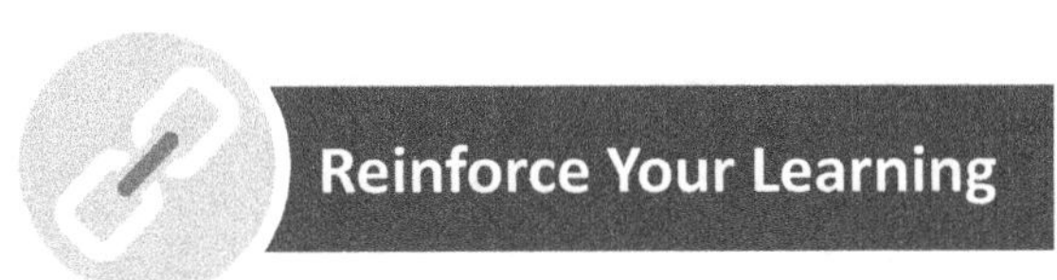

Develop a Work Breakdown Structure. Pick a work-related project such as organizing a celebration and segment it into its major activities. Then list the sub-activities under each major activity. Imagine an outline format, or you could make a chart like the wedding example at the beginning of this concept.

Project: ____

*Project management is the art of creating the illusion that any outcome
is the result of a series of predetermined, deliberate acts when, in fact,
it was dumb luck.*

—Harold Kerzner

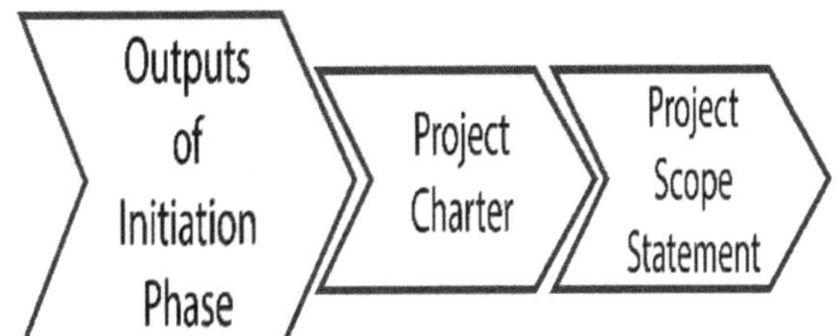

Summary

A Final Word: *Defining Project Scope*

You have completed a chapter on Defining Project Scope.

Information you reviewed in this chapter include:

- Conducting a stakeholder analysis

- Understanding the benefits and obstacle in defining a project scope

- Recognizing common causes of project failure

- Evaluating the critical factors associated with starting projects and your readiness

- Developing a Work Breakdown Structure (WBS)

In the next *Mastering Project Management: Planning for Performance* chapter, we will focus on Planning and Scheduling Projects.

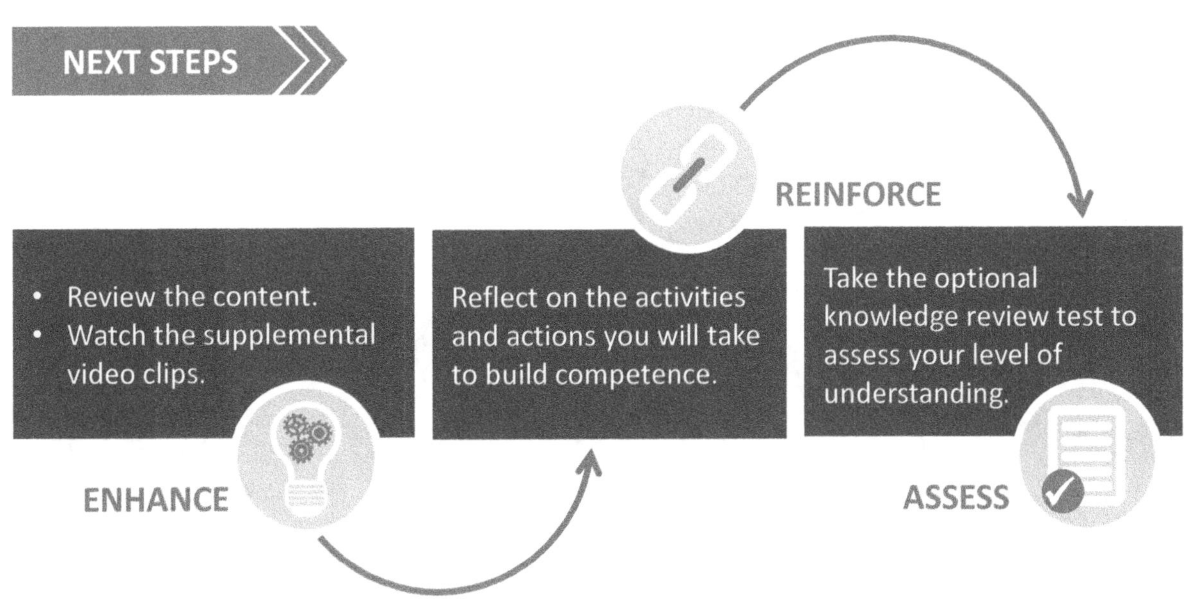

Notes:

Defining Project Scope—Recap Checklist

When it comes to the successful completion of projects, what you do not know can hurt you. Doing the work might be 90 percent of the job, but the initial planning work will impact the whole project. A project well done is always a project well begun. The project scope describes what is in the project and what is not.

1. Identify Project Stakeholders
- ☐ Identify the stakeholders associated with your project.
- ☐ Create a Stakeholder Analysis Matrix.
- ☐ Assess the level of interest and engagement each stakeholder will have in the project and their potential impact on the project.

2. Determine Stakeholder Requirements
- ☐ Identify stakeholders and their requirements early in the project.
- ☐ Ask people who will do the work for their input.
- ☐ Provide opportunities for stakeholders to be engaged in the project.

3. Specify Project Deliverables
- ☐ List the documents that will be used in most projects.
- ☐ Establish channels of communication for the exchange of project information.
- ☐ Communicate progress, obstacles, and other pertinent information on an ongoing basis as appropriate.

4. Draft Project Charter
- ☐ Determine what level of authority you have to initiate projects in your organization.
- ☐ Discuss whether a quote, purchase order, or statement of work (SOW) is sufficient authorization to initiate projects.
- ☐ Identify who in your organization has the authority to sponsor or authorize larger projects.

5. Learn Common Causes of Project Failure
- ☐ Recognize that people's actions may account for the majority of project problems.
- ☐ Consider if you are getting or have all the project requirements.
- ☐ Consider the risks for your project and what could go wrong, and the impact on the project.

6. Prepare for Project Success
- ☐ Establish and agree on project boundaries before starting a project.
- ☐ Document project expectations for cost, scope, timing, and quality.
- ☐ Identify the actions you should take before starting a project.

7. Define Project Scope
- ☐ Meet with stakeholders often to ensure you understand their requirements.
- ☐ Communicate your understanding of the deliverables to the stakeholders.
- ☐ Resolve disagreement among stakeholders early in the planning phase and agree on the quality criteria for project success.

8. Review PMI's Scope Management Model
- ☐ Review PMI's Scope Management Model and associated concepts.
- ☐ Recognize that many project activities and tasks overlap with interdependencies and that project life cycles can vary.
- ☐ Investigate how Agile and SCRUM techniques may help you manage a project.

9. Understand Why Pre-planning is Critical
- ☐ Be aware of what can go wrong as a result of poor project planning.
- ☐ Ensure that the project planning process meets client/stakeholder expectations and requirements.
- ☐ Recognize that careful project planning is the best defense against litigation.

10. Develop Work Breakdown Structure
- ☐ Identify project activities and alternatives by conducting a brainstorming session.
- ☐ Prepare a WBS that defines all the deliverables and work in the project.
- ☐ Break each deliverable into sub-deliverables. The number of sub-deliverables depends on the complexity of the higher-level task.

Action Planning **Competency #1** >>

Planning and Evaluation

Establishes policies, guidelines, plans, and priorities; plans and coordinates with others; aligns required resources; monitors progress and evaluates outcomes; improves organizational efficiency and effectiveness.

Briefly describe how improvement in this competency will help you achieve important results or better meet your job responsibilities.

List courses, books, and independent study opportunities that could help you develop this competency.

Identify one or more people who could help you, either as a role model or source of
information. Write any questions you want to ask each person

135

What specific steps will you take?   **Start Date**   **Finished**

Action Planning | **Competency #2** >>

Resource Management

Demonstrates awareness of technical resources; knows how to apply resources to achieve desired outcomes.

Briefly describe how improvement in this competency will help you achieve important results or better meet your job responsibilities.

List courses, books, and independent study opportunities that could help you develop this competency.

Action Planning | **Competency #2** >>

Identify one or more people who could help you, either as a role model or source of information. Write any questions you want to ask each person

What specific steps will you take? **Start Date** **Finished**

Communication

Makes clear and effective presentations to individuals and groups; listens to others; communicates effectively in writing; can critically review and comprehend information written by others.

Briefly describe how improvement in this competency will help you achieve important results or better meet your job responsibilities.

List courses, books, and independent study opportunities that could help you develop this competency.

Identify one or more people who could help you, either as a role model or source of information. Write any questions you want to ask each person

What specific steps will you take?   **Start Date**   **Finished**

Management is, above all, a practice where art, science and craft meet.

—Henry Mintzberg

Defining Project Scope

Knowledge Review Test

Part A. Knowledge Review Test—Questions

Part B. Knowledge Review Test—Answer Sheet

Part A. Knowledge Review Test—Questions

Defining Project Scope

1. Which of the following is not part of the stakeholder management effort?

    A.   Give them extras.
    B.   Identify them.
    C.   Determine their needs and expectations.
    D.   Manage their expectations.

2. It is important that key stakeholders are as active about the project as you had anticipated. There could be reasons for their lack of engagement which may include:

    A.   Unaware of project activities occurring.
    B.   Against selected project actions.
    C.   Neutral as to what is occurring in a project.
    D.   Satisfied with level of activity.
    E.   All the answers are correct.

3. You can communicate with stakeholders and collect their requirements in a variety of ways, including:

    A.   Interview.
    B.   Focus groups.
    C.   Surveys and questionnaires.
    D.   Direct observation.
    E.   All the answers are correct.

4. Which statement about stakeholder expectations is not true?

    A.   The product or service should be usable as intended.
    B.   Contractual provisions, including schedule and performance standards, should be met.
    C.   Changes should be made immediately, and all stakeholder wants should be satisfied without bureaucratic hassle.
    D.   The seller should assume the responsibility of understanding stakeholder needs and wants and address them effectively.

5. Which of the following documents is NOT typically used in most projects?

    A.   Work breakdown structure (WBS).
    B.   Timesheet report.
    C.   Risk management plan.
    D.   Organization's annual report.
    E.   Lessons Learned.

6. Goals and objectives are correctly stated as "SMART" if they follow this format

    A.   Measurable and time specific.
    B.   Specific, measurable, and realistic.
    C.   Specific, measurable, achievable, realistic, and timely.
    D.   Timely, measurable, verifiable, and achievable.
    E.   None of the answers are correct.

7. A project manager is employed by an organization and is responsible for the furnishing of the completed facility. One of the first things that the project manager for this project should do is to prepare a

    A.   Work breakdown structure.
    B.   Budget baseline.
    C.   Project charter.
    D.   Project plan.

8. A project charter formally authorizes the work to be done and is especially useful for larger projects. For smaller projects what may suffice?

    A.   Formal quotation.
    B.   Purchase order.
    C.   Statement of Work (SOW).
    D.   All the answers are correct.

9. One of the biggest reasons that projects fail is:

    A.   Resources untrained and misaligned.
    B.   Changes in an organization's priorities.
    C.   Tools and systems outdated and inaccessible.
    D.   Assumptions poorly defined and unsubstantiated.
    E.   Team collaboration inadequate and team conflicts.

10. People's actions make up about what percent of the risks in a typical project's Risk Management Plan?

    A.   20 %
    B.   50 %
    C.   70 %
    D.   90 %
    E.   100 %

11. Each project phase is marked by the completion of one or more deliverables. The deliverable for the initiation phase of the project is:

      A.   Project plan.
      B.   Lessons learned document.
      C.   Project charter and scope statement.
      D.   Resource spreadsheet.

12. Ideally, the project manager should be selected and assigned:

      A.   During the initiating phase or process.
      B.   During the project planning process.
      C.   At the end of the concept phase of the project life cycle.
      D.   Prior to the beginning and the development phase of the project life cycle.

13. The document that describes the objectives, work content, deliverables, and an end product of a project is the:

      A.   Project charter.
      B.   Product description.
      C.   Scope statement.
      D.   WBS.

14. Adding extra features or components to a project beyond the original scope requirements is referred to as:

      A.   Quality enhancement.
      B.   Exceeding customer expectations.
      C.   Gold plating.
      D.   Excellent customer service.
      E.   Scope enhancement.

15. The PMI Scope Management Model involves which of the following steps?

      A.   Initiation.
      B.   Planning.
      C.   Definition.
      D.   Verification.
      E.   Change Control.
      F.   All of the answers are correct.

16. Agile project management is characterized by:

    A.   Detailing project deliverables as manageable components.
    B.   Overlapping of many project activities and tasks.
    C.   Monitoring and managing changes to the original project scope.
    D.   Frequent reassessment and adaptation of plans.

17. Project management professionals refer to the critical first 10 percent as:

    A.   The most difficult time in a project.
    B.   The identification of the ten most critical elements of a project.
    C.   Planning the first ten percent of a project.
    D.   The time spent planning a project from start to finish.

18. You are a project manager for a new building project and have just received a project charter. You can create a detailed project schedule only after?

    A.   A budget is created.
    B.   A WBS is crested.
    C.   A project control plan is created.
    D.   A project plan is created.

19. What is a "Work Breakdown Structure"?

    A.   A list of deliverables.
    B.   A map of the entire project.
    C.   A list of cost accounts.
    D.   A hierarchical structure.
    E.   All the answers are correct.

20. A project is plagued by changes to the project charter. The person with primary responsibility to decide what changes are necessary is:

    A.   Project manager.
    B.   Team leader.
    C.   Management.
    D.   Stakeholder.
    E.   All the answers are correct.

Part B. Knowledge Review Test—Answer Sheet

1.   _X_ A   ___ B   ___ C   ___ D   ___ E

2.   ___ A   ___ B   ___ C   ___ D   _X_ E

3.   ___ A   ___ B   ___ C   ___ D   _X_ E

4.   ___ A   ___ B   _X_ C   ___ D   ___ E

5.   ___ A   ___ B   ___ C   _X_ D   ___ E

6.   ___ A   ___ B   _X_ C   ___ D   ___ E

7.   ___ A   ___ B   _X_ C   ___ D   ___ E

8.   ___ A   ___ B   ___ C   _X_ D   ___ E

9.   ___ A   _X_ B   ___ C   ___ D   ___ E

10.   ___ A   ___ B   ___ C   _X_ D   ___ E

11.   ___ A   ___ B   _X_ C   ___ D   ___ E

12.   _X_ A   ___ B   ___ C   ___ D   ___ E

13.   ___ A   ___ B   _X_ C   ___ D   ___ E

14.   ___ A   ___ B   _X_ C   ___ D   ___ E

15.   ___ A   ___ B   ___ C   ___ D   _X_ E

16.   ___ A   ___ B   ___ C   _X_ D   ___ E

17.   ___ A   ___ B   ___ C   _X_ D   ___ E

18.   ___ A   _X_ B   ___ C   ___ D   ___ E

19.   ___ A   ___ B   ___ C   ___ D   _X_ E

20.   ___ A   ___ B   _X_ C   ___ D   ___ E

Chapter 3

Planning and Scheduling Projects

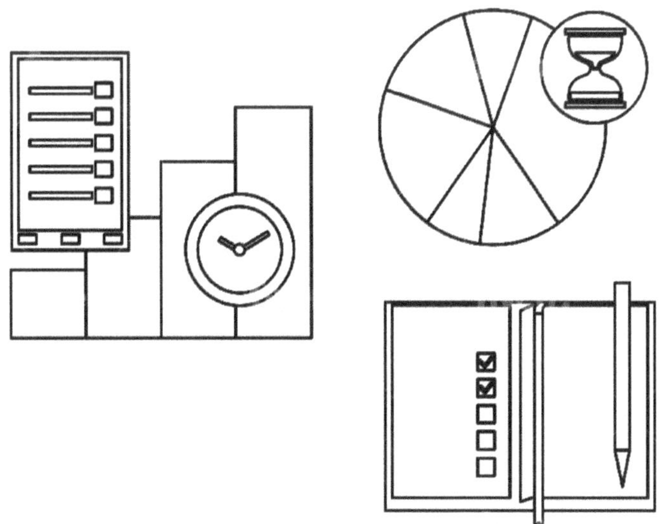

148

A goal without a plan is just a wish.

—Antoine de Saint-Exupéry

Overview

Planning and Scheduling Projects

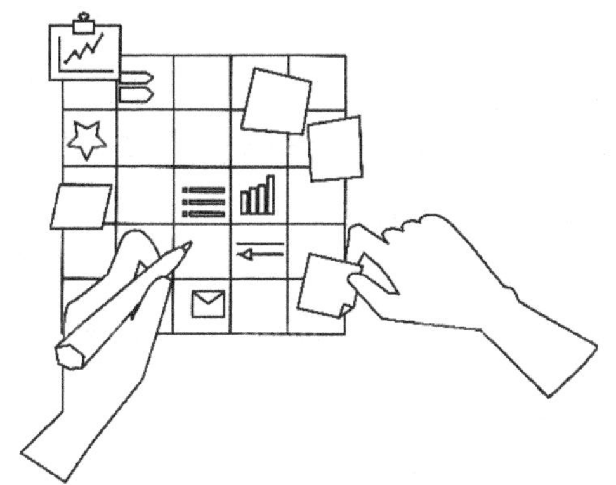

"Strategy without tactics is the slowest route to victory. Tactics without strategy is the noise before defeat."

—Sun Tzu c. 490 BC

Your project plan guides you in managing project costs, resources, schedules, and risks. Thus, effective planning is key to success.

The Project Management Institute (PMI)® produces an annual global survey about portfolio, program, and project management. *PULSE OF THE PROFESSION* ®: Capturing the Value of Project Management, 2015, included the results of this question: In your estimation, what percentage of the projects completed within your organization in the past twelve months **finished within their initial scheduled times**? The answer was **50 percent.** This means that half of all projects take longer than expected. You can read more about the survey results at Pulse (PMI, 2015).

Planning requires that you identify all project activities and tasks, sequence them correctly, and estimate schedules and resources accurately. Be aware that planning is an iterative process. Throughout a project, you will revise your plan as more details become available, tasks are refined, and project changes are approved.

This chapter introduces tools and techniques that help you plan and schedule projects. It also reviews the skills needed to develop and maintain plans, and it discusses common pitfalls. In this chapter, you will learn how to:

- Describe what is in a Schedule Management Plan
- Recognize commonly used project scheduling tools and their applications
- Develop a project schedule
- Explain what is meant by a schedule baseline
- Avoid planning and scheduling pitfalls

The competencies associated with this chapter include:

Planning and Evaluation   **Resource Management**   **Problem Solving**

Take Your Temperature for Utilizing Project Planning and Scheduling Projects
The following assessment will help you identify what you know about project planning and scheduling tools and the areas where you can improve. With you and your organization in mind, read each statement below, and consider how much you agree with it. Use the numbers from 1 to 10, where **1** means you **strongly disagree** and **10** means you **strongly agree**.

DISAGREE 1 2 3 4 5 6 7 8 9 10 **AGREE**

______ 1. I can outline components for building a project schedule and baseline.

______ 2. I know how to sequence project tasks.

______ 3. I understand how to determine the effort needed for tasks.

______ 4. I am confident in my ability to set project milestones.

______ 5. I can develop Gantt charts.

______ 6. I can schedule activities using a spreadsheet.

______ 7. I know how to develop a project network diagram.

______ 8. I can calculate a project's critical path.

______ 9. I can describe the Program Evaluation and Review Technique (PERT).

______ 10. I know how to avoid common project planning and scheduling pitfalls.

______ **Total**

Take a moment to reflect on the following before you move forward in the chapter:

What problems or situations have you experienced or observed related to planning and scheduling projects at work?

Take a few minutes to reflect on your self-assessment. List two or three areas you want to improve..

 As you progress through this chapter consider what actions you can take to demonstrate competence in the following three competency areas:

1. **Planning and Evaluation.** Establishes policies, guidelines, plans, and priorities; plans and coordinates with others; aligns required resources; monitors progress and evaluates outcomes; improves organizational efficiency and effectiveness.

2. **Resource Management.** Demonstrates awareness of technical resources; knows how to apply resources to achieve desired outcomes.

2. **Problem Solving.** Recognizes and defines problems; analyzes relevant information; encourages alternative solutions, develops plans to solve problems.

Build the Project Schedule

After you initiate a project, define the scope, and start assembling the project's work activities, it is time to develop the project schedule. Scheduling, along with determining the level of effort for tasks and activities, is one component of a project management work plan.

The **Project Life Cycle** illustration below shows the phases involved in a project life cycle, the graduated use of resources, and the project outputs. This course focuses on the **planning** phase.

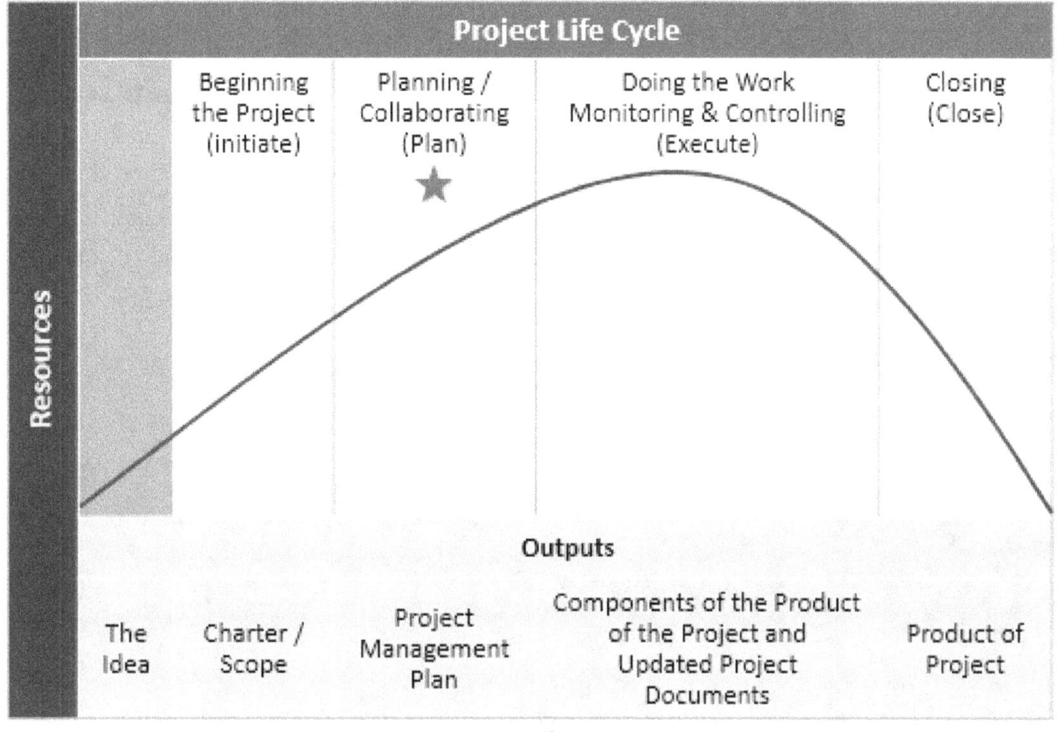

A ***Project Management Plan,*** often referred to simply as the ***Project Plan,*** addresses how tasks and resources will be planned, executed, monitored, and controlled.

The initial project management plan is referred to as the ***baseline schedule and plan****.* A *baseline* is a fixed schedule that represents the standard used to measure the performance of a project.

In project management, there are **three baselines—schedule** baseline, **cost** baseline, and **scope** baseline. The combination of all three baselines is often referred to as the **performance measurement baseline**.

The Project Management Plan Integrates Knowledge Areas

The illustration below shows how the Project Management Plan enables integration among a project's knowledge areas.

This course focuses on the **schedule knowledge area**.

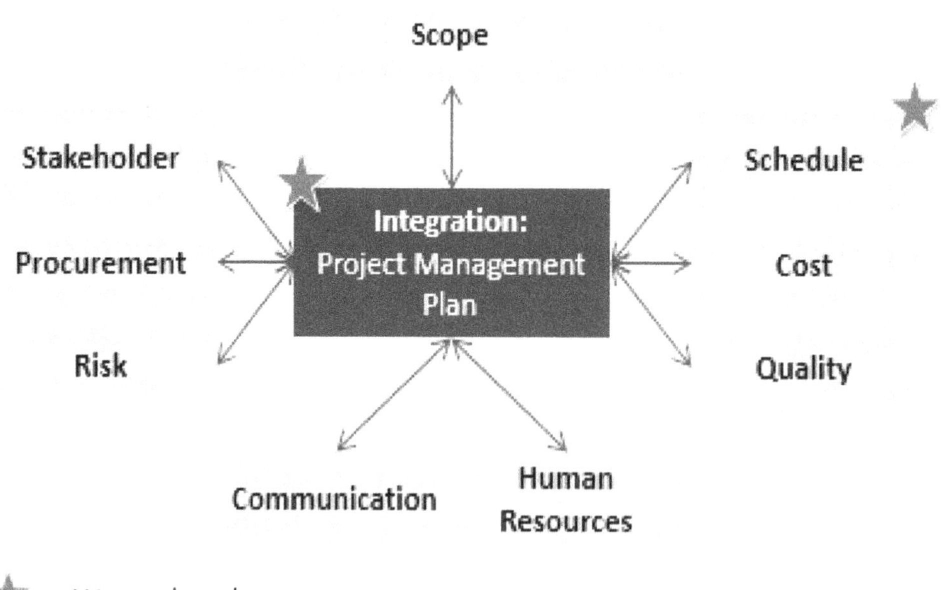

The Project Planning Phase

The following illustration shows some of the major steps and project deliverables in the planning phase.

Phase	Description/Purpose	Deliverables/Tools
Plan: Plan from beginning to end of the project	• Define tasks, activities, inter-dependencies, and activity times • Refine resource requirements • Develop work plan, milestones, and schedule • Define project control process, standards, and procedures • Develop risk management plan • Identify communication needs • Refine and manage expectations	• Schedule Management Plan • Work Breakdown Structure • Status Report Format • Time/Schedule -Tracking Format • Risk Analysis and Management Plan • Stakeholder Communication Matrix • Communication Plan • Project Budget • Benefit-Cost Analysis

Components Needed for Building a Schedule

The components or inputs for planning and developing a schedule include several kinds of information, including the following:

- Activity list and attributes
- Resource list and calendar
- Activity duration estimates
- Project scope statement
- Constraints and risk assessment
- Project staff selection and assignments

A Project Schedule is a Formal Agreement

It is useful to think of the project schedule as a kind of contract. It is a formal agreement between the client/stakeholders and the project manager and team.

Effort and resources are needed to create a schedule, but the expenditures are worthwhile because the schedule is intended to provide a document that accurately reflects what the project team is capable of delivering, and it considers the schedule, cost, and scope as specified in the Project Management Plan.

Changes Will Happen as the Project Progresses

Changes occur in almost all projects. The larger and more complex the project, the more confident you can be that changes will occur; some may be significant.

Baseline revisions show the impact of the changes on the original schedule, cost, and scope.

A periodic review of the *original scope* or *scope baseline* includes a comparison of the original scope of the project plus or minus any scope changes. In other words, a comparison allows you to see the gap between the baseline initially agreed to and any revised baseline.

Project managers must show clients/stakeholders the impact a change may have on a project *in advance* of actually making the change. That way, the client/stakeholders can weigh the disadvantages (schedule, cost, or scope) against the benefits of including the changes in the project. Using this information, they can also make an informed cost-benefit analysis of the proposed change, if needed.

How to Select Planning and Scheduling Tools

Numerous methods/approaches and tools are available to assist you in planning and scheduling the activities defined in a Work Breakdown Structure (WBS).

Some of the most useful include a Gantt chart, milestone chart, the deterministic (known relationships) approach or network, and the probabilistic approach or network.

For small projects, a Gantt/Bar Chart and a Milestone Chart are sufficient. For large projects, project managers use a combination of all four tools.

The following chart reviews the criteria to consider when you evaluate and select tools for a project and provides a high/low assessment of the methods/approaches listed above.

Criteria for Selecting Scheduling Tools (*** = High   * = Low)**

	Gantt/Bar Chart	Milestone Chart	Small Project Tool	Network Approaches	
				Deterministic /Critical Path (CPM)	Probabilistic/ Program Evaluation Review Technique (PERT)
Ease of preparation	*****	*****	***	**	*
Ease of updating	*****	*****	**	*	*
Use with simple projects	*****	*****	*****	*	*
Use with complex projects	*****	*****	**	*****	*****
Useful as a communication tool	*****	*****	***	***	**

What to do:

- ☐ Identify project methods/approaches and tools you are familiar with and those to learn more about.
- ☐ Consider a baseline schedule as a formal agreement.
- ☐ Review the baseline schedule to validate that the project team can provide the deliverables as scheduled.

Other actions:

- ☐ _________________________________
- ☐ _________________________________
- ☐ _________________________________
- ☐ _________________________________

✓ Various approaches and tools are available to help you plan a project. Project managers should select the ones they are comfortable using and that best serve their project and project team needs.
✓ A published project schedule is a baseline. It is what you will be measured against, and you should consider it to be a type of contractual or formal agreement.
✓ Change controls are the control mechanism to bridge the gap between the original baseline schedule and the actual schedule.
✓ When the schedule changes, all associated project documents must be updated to ensure appropriate project communication and that the project is monitored and controlled appropriately.
✓ The initial project management plan is referred to as the baseline schedule and plan.

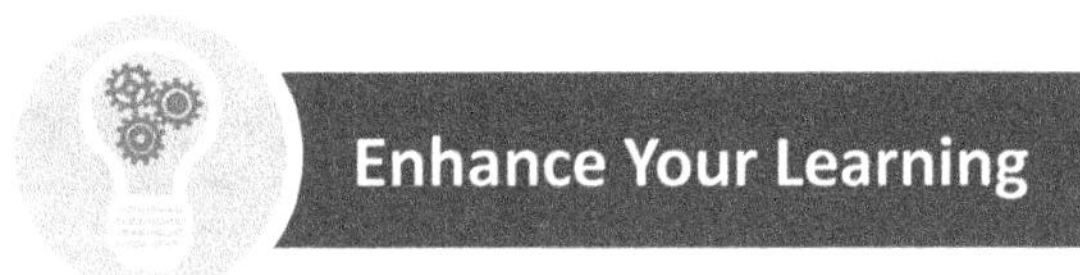

Watch the following 3-minute video to learn more about project scheduling from ProjectManager.com.

		Available at:
Bridges, J. (2018). *What is Project Scheduling?*		https://www.youtube.com/watch?v=WNWSQOynrl0

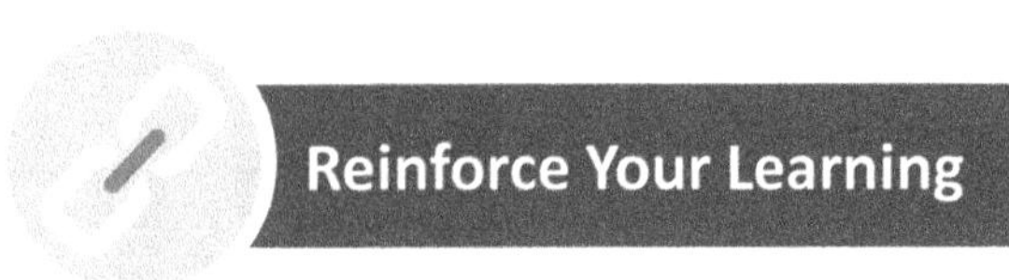

Planning and Scheduling Challenges. List three planning and scheduling challenges and identify methods or tools that could help improve the situation.

Planning and Scheduling Challenges	Helpful Project Planning Tools
1.	
2.	
3.	

Sequence Project Tasks and Activities

People sometimes use task and activity interchangeably; however, there are differences. When managing a project, task refers to a unit of work to be done. Activity refers to an event or happening. According to the Project Management Institute's *PMBOK ® Guide,* an activity is a distinct, scheduled segment of work. Stated differently, a task is the "what," and an activity is the "how and when."

Project managers construct project schedules based on practical methods, or "rules of thumb," that are not guaranteed to be optimal or perfect but are useful.

Rules of thumb come from a variety of sources. Some examples include project manager professional judgment; the experience of subject matter experts (SMEs), and ideas from the project team and stakeholders.

Project manager judgment is based on knowledge and experience. For example, knowledge may come from studying for the certifications offered by PMI. The prior experience might include having worked on projects like the current project or referring to a previous schedule management plan. Project managers may also be fortunate enough to have some documented "lessons learned" from an earlier project.

SMEs (Subject Matter Experts) may be in the best position to identify and sequence current activities and work packages.

Project **meetings** are generally an excellent place to discuss ideas and conduct brainstorming sessions because the meetings typically include knowledgeable team members, SMEs, stakeholders, and others who may be affected by the project.

One goal of sequencing project tasks and activities is to identify the logical sequence. This includes the relationship between tasks, such as "complete A before B can start." Another goal is to identify any constraints.

PMBOK is a registered mark of the Project Management Institute, Inc.

The *four types of dependencies* are: mandatory, discretionary, external, and internal. *Mandatory* dependencies may occur because of physical limitations. A successor activity may not physically be able to finish until another predecessor activity is complete.

Legal constraints may also require a particular sequence of activities.

What to do:	**Other actions:**

What to do:

- ☐ Review a sequence of project activities and validate they are accurate.
- ☐ Pay attention to how one activity relates to another (predecessor and successor).
- ☐ Inform stakeholders that you reviewed the sequence of activities and that you agree or want to recommend changes.

Other actions:

- ☐ ______________________________
- ☐ ______________________________
- ☐ ______________________________
- ☐ ______________________________
- ☐ ______________________________

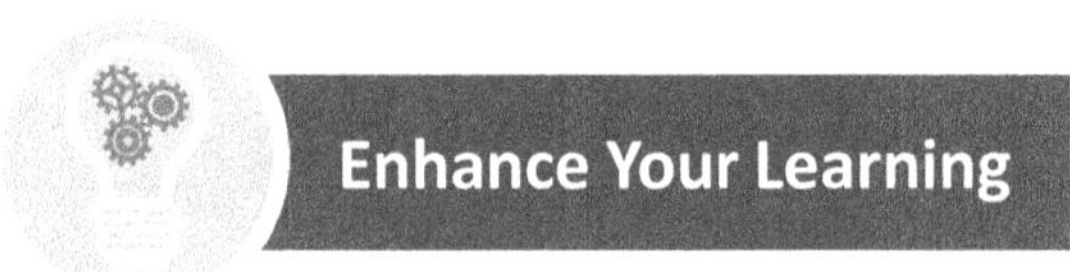
Remember

- ✓ You must plan/schedule the sequence of activities so that you have the right resources available at the appropriate times.
- ✓ Precedence diagramming is a tool project managers use to determine the type of relationship between two activities.
- ✓ The four major types of dependencies are: mandatory, discretionary, external, and internal.

Enhance Your Learning

View the following 6-minute video from Andy Kaufman about activity sequencing:. The project example in the video assumes he is going to take a family trip to Florida to watch a space launch.

Kaufman, A. (2015). *Sequence Activities Demonstration.*		Available at: https://www.youtube.com/watch?v=H-1Ab30_rrM

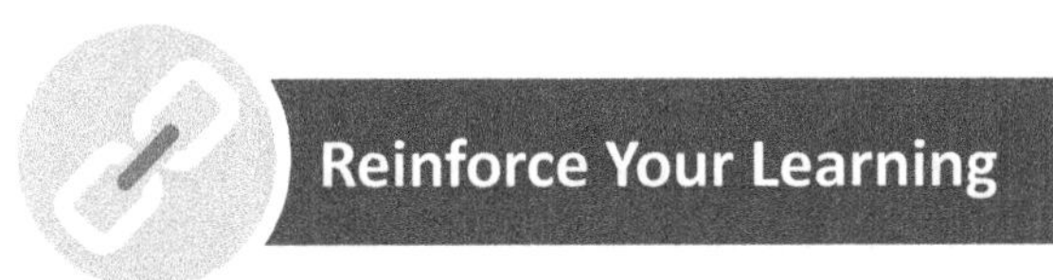

Activities Template. Use the template below to list activities involved in making copies.

Based on your experience as a subject matter expert, list all the activities required to complete the work of making copies. For example, the first activity would be to gather the documents you want to copy, which is shown as the first activity below.

For each activity, identify the predecessor activity and the successor activity, and any constraints. After completing your activity list, review it with a colleague.

Project Name: Making Copies

Activity List

ID#	Activity Name and Description	Predecessor Activity	Successor Activity	Constraints
1	Gather documents for hard copy.			

By failing to prepare, you are preparing to fail..

—Benjamin Franklin

Determine Effort for Tasks

Effort often refers to the number of labor units or man-hours needed to complete a task or activity. However, *effort* can also refer to human resources, material, or financial project needs.

In this concept, we will focus on estimating and planning the time and resource effort required to complete project work.

Work Packages

Recall, as part of developing a WBS, you will define work packages. A rule of thumb is no work packet is more than 80 hours of work or less than 8 hours.

The following is the information you will define for each work package:
- Work to be done (what)
- Time to complete a work package (how long)
- Time-phased budget to complete a work package (cost)
- Resources needed to complete a work package (how much)
- A single person responsible for the unit of work (who)
- Monitoring points for measuring progress

Planning Techniques

For complex projects, a WBS requires thought and planning to determine the realistic effort required to complete work. Two ways of evaluating work effort are work decomposition and what is called the rolling wave or progressive elaboration method.

Decomposition means taking a deliverable, task, activity, process, or event and breaking it down into smaller, more manageable components to provide better control.

The rolling wave or ***progressive elaboration*** method means walking through the activities or work packages over many iterations. Each iteration may identify something previously missed, refine what was determined earlier, or validate that all work packages are accounted for. This is a litmus test for the "no rocks left unturned" approach.

Scheduling Tools and Techniques

Here are some of tools and techniques for estimating the time and resource effort needed on a project:

- Expert judgment
- Alternative analysis
- Published estimating data
- Bottom-up estimating
- Group decision-making
- Reserve analysis
- Deterministic scheduling approach
- Probabilistic scheduling approach

Expert judgment can come from people with experience in the kind of work to be estimated. Experienced people can help predict the time and resources needed to do a piece of work, help pinpoint the necessary skills, and identify people who may have those skills.

Alternative analysis considers ways the work could be done. For example, an activity planned initially to be done manually, such as a data collection task, might be automated. That way, a lesser-skilled resource may have the ability to work on activities originally assigned to someone highly skilled.

Other alternatives include types or sizes of machines, varying output rates, or make-rent-or-buy considerations. As an example, depending on the need for equipment and the lead time needed to purchase it, a rental approach may save time and money.

Published estimating data is one source of information for use in bottom-up estimating. It can come from a variety of sources, for example, production rates, unit costs, material costs, and equipment costs from across the globe.

No matter the source, you must consider how the data relate to your project's environment. For example, you do not want to use data that are no longer accurate/up to date or that do not closely reflect the current situation.

Bottom-up estimating leverages the experience of the project's SMEs and others knowledgeable about the activities. Working knowledge is instrumental in identifying not only the effort required to complete work, but also any dependencies between work units. Each work unit is estimated individually, and then all the individual estimates are added together to arrive at a total.

This contrasts with a *top-down approach*, where the completion date may be imposed on the project team. Bottom-up is considered the most accurate method in estimating cost and duration. However, be aware that it requires the most time.

Group decision-making techniques include the concepts of multiple perspectives, knowledge areas, people close to the processes, and buy-in of the capability to meet the estimates. It is essential that project team members and stakeholders are aware of what is being committed to from a schedule standpoint. They may have input that could change projections, and the sooner you know that the better.

Reserve analysis involves the concept of including a "buffer" reserve to account for estimate uncertainties. They can be referred to as contingency reserves or time reserves. The reserve analysis is based on the level of confidence the estimators have in their work, and it considers the similarity with prior experiences. It is important for reserves to be included in the schedule and identified as such. It is also helpful to have organizational standards as to how reserves should be handled for consistency across projects.

Deterministic scheduling approach. If you have confidence in the information and data that is being estimated and used in scheduling, a deterministic scheduling approach may be useful. In this approach, you know what outcomes are precisely determined through known relationships. In such methods, a given input will always produce the same output. Stated differently, project schedules are networks of tasks, activities, or resources connected to each other with known dependencies that precisely and accurately describe the tasks, activities, or resources.

Probabilistic or three-point scheduling approach. If you have developed a project schedule or budget, you know it is often difficult to make accurate estimates. The best strategy is to estimate the individual times and costs and then add them together to get the total time and cost. Doing this will also help you in tracking progress and monitoring costs. However, this can be difficult. Even under the best conditions, it may be impossible to get a realistic schedule.

Experience has shown that you can improve the accuracy of an estimate by using the **probabilistic** or **three-point estimating** approach. This approach has been proven in a wide range of projects. It considers the **best-case** scenario and the **worst-case** scenario and balances them with the **most likely** scenario.

Making Probabilistic or Three-Point Scheduling Estimates

The following formula is commonly used in making probabilistic estimates (**E**):

$$E = (O + 4M + P)/6$$

(O) is the best-case scenario/most optimistic estimate, added to four times the most likely estimate (**M**). Add the worst-case scenario/most pessimistic estimate (**P**). Then divide that number by 6. This approach is equally valid for both the schedule and the budget.

The following are guidelines for using the probabilistic approach:

- Consider it when you do not have better information. If better information is available from previous projects or from input others may give, use that information.
- Confirm the best case and worst-case scenarios with people who have experience in the same type of project. This will help improve the accuracy of your estimates.
- Combine these time estimates with your network diagram to form the basis of your schedule.
- When needed, use this same approach to estimate costs and develop a budget.

Complete an Activity List with Resources and Time Estimates Chart

Use the outputs from an activity duration estimation process to create an Activity List with Resources and Time to Complete chart. It lists the activities, resources (who), and the time to complete (duration) estimates, along with any constraints.

Activity List w/ Resources and Time to Complete Estimates

ID#	Activity Name	Resources (Who?)	Predecessor Activity	Successor Activity	Time to Complete	Constraints

After you identify the activities, sequence them, add resources, and provide an estimated duration, you will be ready to develop a baselines schedule for the project.

In the next concept, we will discuss how to convert an activities list into a project work schedule.

What to do:	**Other actions:**
☐ Identify your best sources of information for determining the effort required to complete activities and tasks.	☐ _____________________
	☐ _____________________
	☐ _____________________
☐ Consider using this same approach to estimate costs and develop a budget.	☐ _____________________
☐ Use the probabilistic or three-point estimating approach when you do not have better information.	☐ _____________________

- ✓ If you have accurate and up to date information, use it instead of the probabilistic approach. The more accurate the information, the better your schedule and budget.
- ✓ Remember that estimates are only estimates. To detect potential issues as early as possible, you must monitor the project as work progresses.
- ✓ Probabilistic or three-point estimating requires that you provide estimates for the least amount of time or effort, the most likely amount of time or effort, and the most amount of time or effort.
- ✓ When your project has work similar to work done in another project, refer to the schedules developed for that project. They may be more accurate than your estimates.

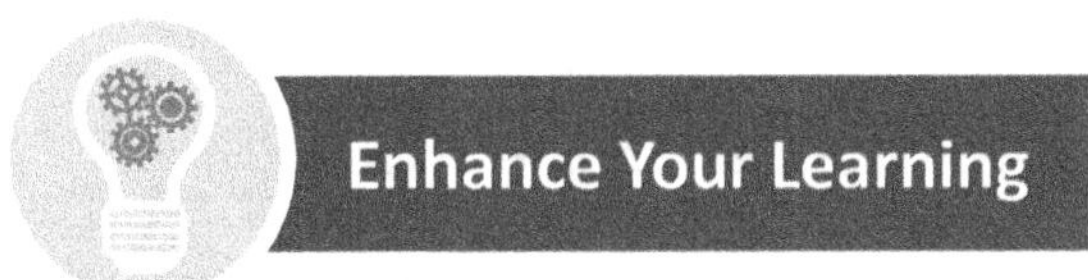

Watch this 2-minute video explaining Level of Effort by Janis.

Janis. (2020). *What is Level of Effort?*		Available at: https://www.youtube.com/watch?v=u5JJOiSIaSI

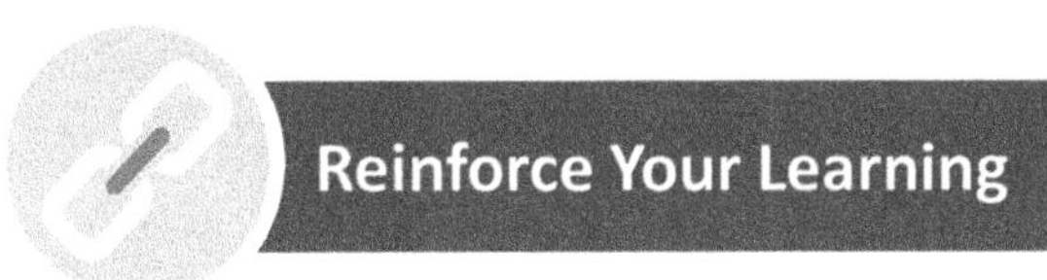

Activity List with Resources and Time Estimates. Return to your activities list for making copies. Using the Activity List with Resources and Time to Complete Estimates template below, identify the resources and estimated time needed to complete each of your copy making activities. You will find a blank worksheet in Appendix "Part A" to copy and use on other projects.

Activity List with Resources and Time to Complete Estimates

ID #	Activity Name	Resources (Who)	Predecessor Activity	Successor Activity	Time to Complete (Duration)	Constraints

If you don't know where you are going, you'll end up someplace else.

—Yogi Berra

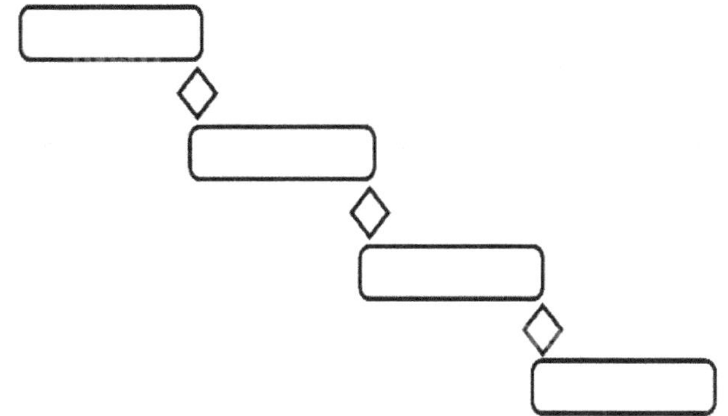

Set Project Milestones

For your project to be successful, you will need to establish challenging yet realistic and achievable milestones. The milestones will help you visualize your project's key points and important junctions.

You can think of **milestones** in several ways and use them according to your project's needs.

A *project milestone* can be:

- A task of zero duration that shows an essential achievement in a project.
- A significant point in the project that defines where you are at any given point, and that is verifiable by others.
- Any point in the project that requires approval **or** a decision before proceeding to the next point.
- Points that represent a clear sequence of events that incrementally build until your project is complete.

How to Develop Milestones

Project managers develop milestones in various ways. Some project managers start by planning forward from the present into the future; some start with the end in mind and work backwards, and some plan in a random way, typically using sticky notes.

No matter the method you choose to plan your project milestones, here are the typical steps you will follow:

1. Define/name your milestones
2. Create links
3. Assign dates
4. Attach tasks
5. Allocate tasks

For larger and more complex projects with multiple components, you may need to determine strategies to link components and develop a series of milestone charts.

Project Milestone Charts

A project milestone chart lists project activities and deadline dates. You can include other information as needed for your project. For example, you might add the team members involved in each activity and completion dates as they occur.

Milestone charts are easy to construct and provide informational highlights for the project. They focus on responsibilities and project status.

The following is a sample milestone chart that shows significant points in a project:

Project Milestone Chart

Task/Activity	By	Deadline	Completed
1. Study	HAJ	9/30	9/15
2. Design	HAJ	10/15	10/15
3. Purchase	DMM	10/31	10/31
4. Contract	DWT	11/15	
5. Test	DMM	11/30	
6. Operate	JRZ	12/15	
7. Accept	WED	12/31	

What to do:

- ☐ Review a milestone chart that lists significant events or decision points with deadline dates.
- ☐ Select major events that incrementally build until your project is complete.
- ☐ Specify what and who needs to approve/decide before proceeding to the next point.

Other actions:

- ☐ _______________________
- ☐ _______________________
- ☐ _______________________
- ☐ _______________________
- ☐ _______________________

- ✓ A project milestone chart consists of a list of activities with deadline dates, and you can include additional information as needed.
- ✓ Milestone charts provide informational highlights.
- ✓ The focus of a milestone chart is on responsibilities and project status.

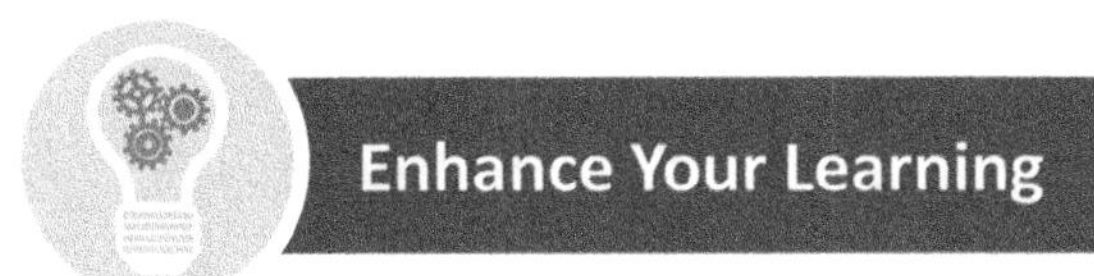

Watch the following 4-minute video by Mike Clayton to learn more about milestone planning.

Clayton, M. (2017). What is Milestone Planning?		Available at: https://www.youtube.com/watch?v=aJcd1rQOZ9w

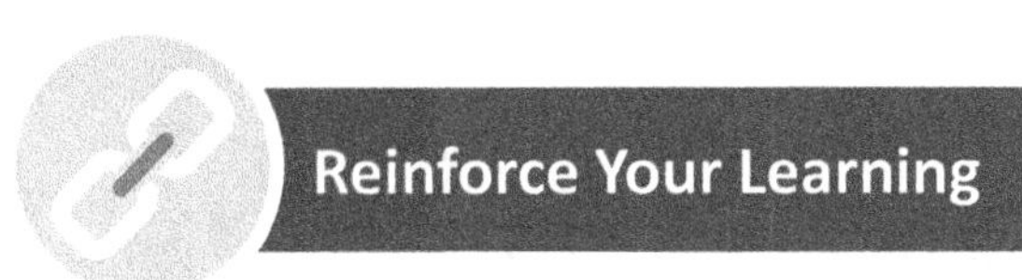

Milestones. Milestones are critical points in a project life cycle. How would you logically re-order the following list of milestones?

- Use it

- Deliver it

- Test it

- Do it

- Prepare it

172

Have a good plan. Execute it violently. Do it today.

—General Douglas McArthur

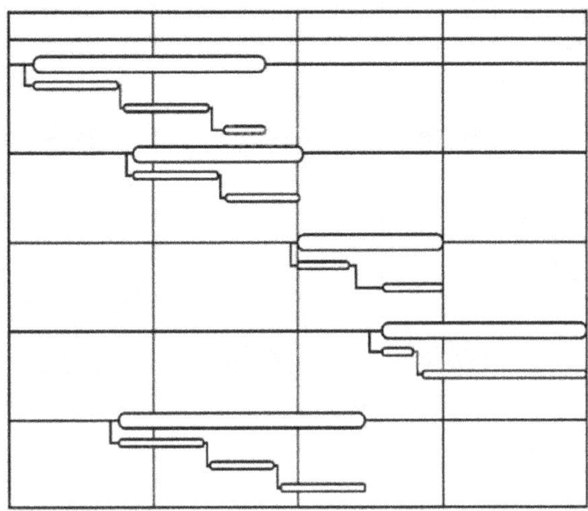

Develop Gantt Charts

Gantt or bar charts are popular because they present an easy-to-understand, clear picture of a project's progress and status. The format is typically a two-dimensional representation of the project schedule, with activities down the rows and time across the columns. This tool gets its name from Henry Gantt, who introduced this way of charting in the early 1900s.

Project managers use Gantt charts throughout a project's life cycle for planning, resource scheduling, and status reporting. Because Gantt charts provide a clear overview of the project schedule, they are often posted on the walls of project offices.

The major **weakness** of Gantt or bar charts is the **absence of dependency** relationships among project activities. If, for example, **float** or **slack** is used on an early activity in the network chain, it cannot be used on a later activity in the same chain, and this dependency is not shown on a bar chart.

Although some computer software will develop bar charts with dependency lines, the lines soon become difficult to read and thus defeat the simplicity of the bar chart.

Note that the bar chart is derived from the project network and **not** the other way around.

You can expand a basic Gantt chart to include other information such as who is responsible and the proposed budget for each task or activity.

Gantt charts can also show actual expenses and times as well as other accumulated information. These charts can be as simple or complex as needed.

Gantt Chart Example

The following **Schedule Management Work Plan** example illustrates how a Gantt or bar chart depicts the time duration for each activity listed. You list activities in the rows and time across the top. For simplicity, the time in this example is represented in months.

Schedule Management Plan

ID#	Activity	1	2	3	4	5	6	7	8	9	10	11	12
1	Gather documents for hard copy	X											

Computer Generated Gantt Chart Example

For complex projects, project management software is useful for making Gantt charts. For example, we made the following Gantt chart for the Making Copies project using Microsoft Project software:

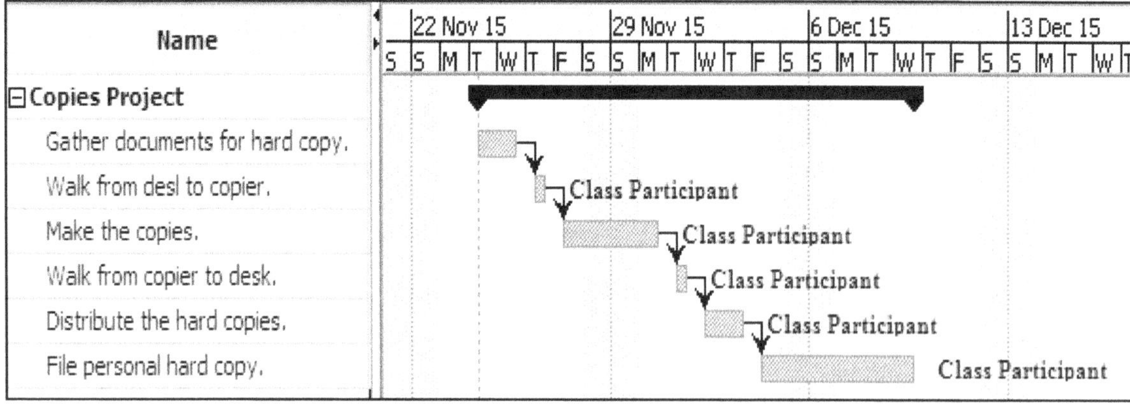

After you finish your Gantt chart, you will have a useful tool to help you manage your project. You can use it to track progress, see when the next activity should start, determine how much money has been spent, and help identify problems such as schedule and cost overruns.

What to do:	**Other actions:**
☐ Select a simple work-related project and create a WBS outlining the major activities or tasks.	☐ ___________________________ ☐ ___________________________
☐ Develop a Gantt or bar chart by listing the major tasks along with start dates and the length of time for each task.	☐ ___________________________ ☐ ___________________________
☐ Try producing a Gantt chart using one of the many project scheduling software tools.	☐ ___________________________

✓ The Gantt chart is a flexible tool that you can use to assign responsibility and track costs, completed (and partially completed) work, overruns, and other project information.

✓ Defining activities is a process you will work on throughout the life of the project in iterations that become more and more detailed.

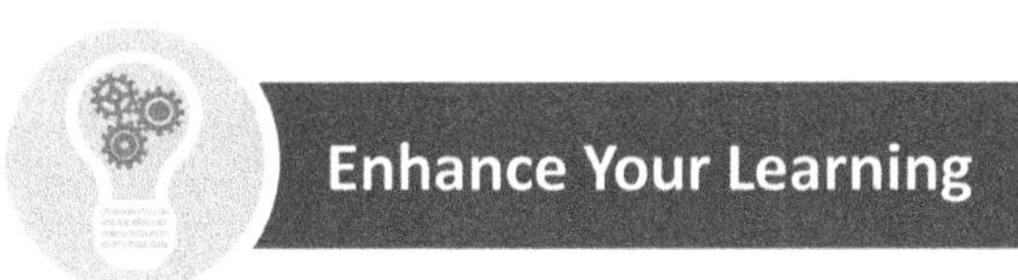

Watch the following 10-minute video on how to make a Gantt Chart by Mike Clayton

Clayton, M. (2022). *How to Create a Gantt Chart in 9 Easy Steps*.		Available at: https://www.youtube.com/watch?v=FX70rWEE8eY

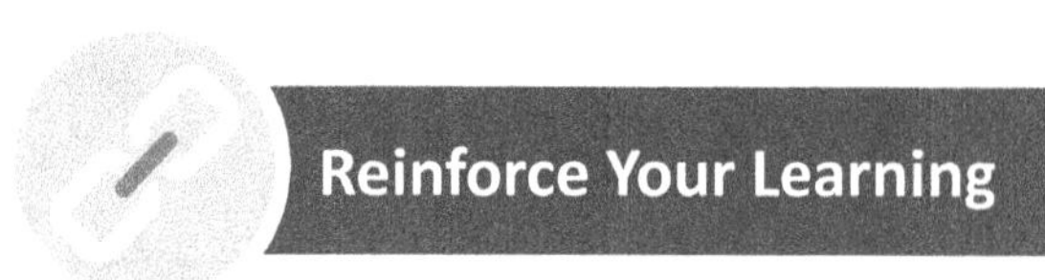

Schedule Management Plan. Assume you are planning a retirement celebration or another project you are familiar with. Develop a Schedule Management Plan using the Gantt chart template that follows. List the project's major activities or tasks, along with start dates and the length of time for each task in months. You will find a Schedule Management Plan worksheet in Appendix Part B to copy and use on other projects.

Schedule Management Plan

Project Name: _______________________________

ID#	Activity	1	2	3	4	5	6	7	8	9	10	11	12

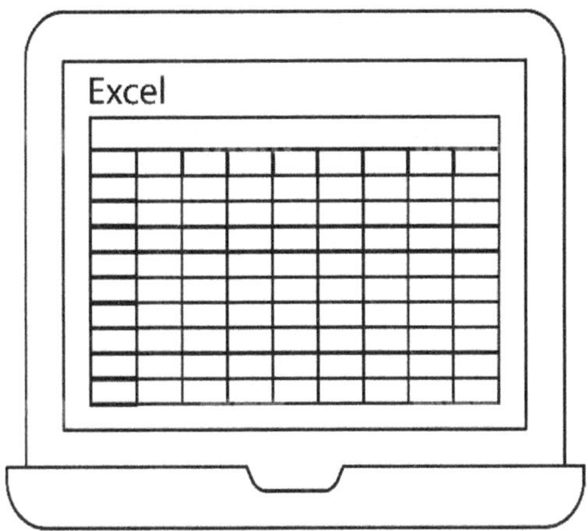

Use a Spreadsheet for Scheduling Small Projects

For small projects that you manage carefully and that have participants who feel part of a team, you can use a spreadsheet to manage the project. This is because you will not need as much detail as you would in a larger project.

As an example, you might manage a small project using an Excel spreadsheet. In this concept, we will review several sample spreadsheets. You can also find many spreadsheet examples and templates online.

Even in a small project, your emphasis is still on the project deliverables, but you can manage them with a simplified WBS matrix. A WBS work package becomes a task activity assigned to one person or one organizational unit.

For this method, you list the task activities on the left side of the spreadsheet, followed by the people/function or work units involved in the project. Then the estimate level of effort time, which is typically tracked monthly in hours, is designated horizontally across the spreadsheet, followed by the same people/function or work units.

Specific individuals/functions or work units track their Level of Effort hours and enter their hours into the spreadsheet monthly.

The following is an example of an interesting "Small Projects Tool" spreadsheet where monthly tracked "Level of Effort" hours are entered into the spreadsheet, first "in aggregate" for the month by the people/function or work units under "Level of Effort per Month" and subsequently segmented separately under the "Level of Effort by Task." The spreadsheet may be programmed to automatically calculate and display the "Level of Effort" total hours on a monthly basis by both the Individual/function or work unit, and by the specific Tasks for the small project.

As an example, the following "Small Projects Tool" spreadsheet was used for a *Schedule and Level of Effort Analysis Chart for a Hypothetical Building Renovation Project*: **Wes's Shop**.

A B C D E F G H I J K L M N O P Q R S T U V W X Y Z AA AB AC AD AE AF AG AH AI AJ

ABC Contractors

Small Projects Tool

Schedule and Level of Effort Analysis Chart

Project: _Wes's Shop_

LEVEL OF EFFORT BY TASK SERIES (IN HOURS)

TASK ACTIVITIES	A. Project Manager	B. Department Head	C. Work Scheduler	D. HR Specialist	E. Project Designer	F. Technician/CAD	G. Purchasing	H. Material Expediter	I. Fabricators	J. Accounting	K. Technical Editor	L. Clerk Typist	M. Shipping Manager	N. Marketing	O.	TOTAL HOURS
01.0 - Project Management	50	2												2		
02.0 - Quote	4	4	4		2					2	2	2		2		
03.0 - Pre-project	4	4	2	2												
04.0 - Contract negotiations	4	4												4		
05.0 - Kickoff	4	4	4	2	4	2	2	2	2					4		
06.0 - Planning/Scheduling	4	2	16		4	4	16	8								
07.0 - Design	4	4			16	16			4							
08.0 - Execution	4	2	4		8	4	4	8	16				4			
09.0 - Shipping/Delivery	2		4				4	8	16	4			8			
10.0 - Closeout	4	2	2	2	2	2				16	2	2	4	4		
11.0 - Warranty Phase	2										2			2		
12.0 - Lessons Learned	2	2	2		2				2					2		

LEVEL OF EFFORT PER MONTH (IN HOURS) — MONTH

	1	2	3	4	5	6	7	8	9	10	11	12	Total
A. Project Manager	8	8	8	8	8	8	8	8	8	8	8		88
B. Department Head		8	4	4	4	2	4		2		2		30
C. Work Scheduler		6	8	8		2	2	4	4	2	2		38
D. HR Specialist		2	2							2			6
E. Project Designer		2	8	4	4	4	4	4	2	4	2		38
F. Technician/CAD		2	2	2	4	4	4	4	2	4			28
G. Purchasing			2	4	4	4	4	2	2	4			26
H. Material Expediter		2	4	4		4	4	4	4				26
I. Fabricators		2		2	2	8	8	8	8		2		40
J. Accounting		2						4	16				22
K. Technical Editor		2								2	2		6
L. Clerk Typist		2								2			4
M. Shipping Manager							2	2	8	4			16
N. Marketing	2	4	6							6	2		20
O.													0
TOTAL HOURS	10	42	44	36	26	36	40	40	56	38	20	0	388

What to do:

- ☐ List the tasks for a small project you are familiar with or have observed.
- ☐ Identify and list individuals/functions or work units responsible for preforming the work.
- ☐ Develop a spreadsheet for your small project.

Other actions:

- ☐ _______________________________
- ☐ _______________________________
- ☐ _______________________________
- ☐ _______________________________
- ☐ _______________________________
- ☐ _______________________________

✓ For small projects, consider using a spreadsheet to identify tasks, assign responsibility, and track costs.
✓ Numerous spreadsheet templates are available online.

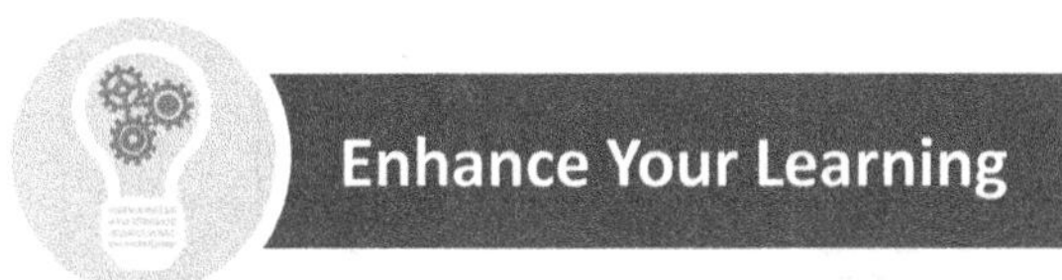

View the following 3-minute video to learn more about five spreadsheet templates.

Bridges, J. (2017). *Top 5 Excel Project Templates - Project Management Training..*		Available at: https://www.youtube.com/watch?v=BAY78mJr9cQ

Also evaluate the Small Projects Tool build in Excel that displays a Schedule and Level of Effort Analysis Chart by using the empty Excel template.

Donahue, W. (2020). *Small Projects Tool - EMPTY SPREADSHEET.*		Available at: https://app.box.com/s/gq1t096g0cs51bbi4dwzi681xutzrh72

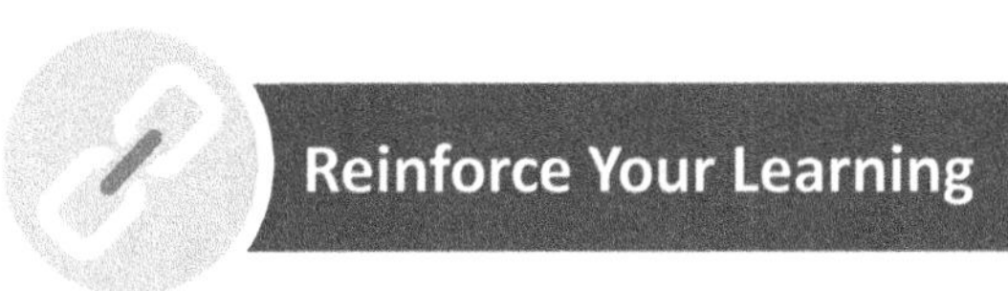

Use a Spreadsheet or Small Projects Tool.
Part 1. For the spreadsheet shown for Wes's Shop, information in the shaded area is missing. Can you fill in the missing information so that all of the rows & columns add up? (Be careful that there is no unique answer.)

Part 2. On your own, use the "Small Projects Tool" spreadsheet worksheet that follows for a simple construction or fabrication project (or project of your choice). Customize the tasks, identify the specific individuals (functions) or organizational units, estimate each one's monthly time in hours by task activity and practice completing the Excel spreadsheet.

Part 1. Wes's Shop

Small Projects Tool
Schedule and Level of Effort Analysis Chart

Project:

LEVEL OF EFFORT BY TASK SERIES (IN HOURS)

TASK SERIES

Schedule portion (MONTH columns 1–12 are blank for all task activities):

TASK ACTIVITIES	1	2	3	4	5	6	7	8	9	10	11	12
01.0 - Project Management												
02.0 - Quote												
03.0 - Pre-project												
04.0 - Contract negotiations												
05.0 - Kickoff												
06.0 - Planning/Scheduling												
07.0 - Design												
08.0 - Execution												
09.0 - Shipping/Delivery												
10.0 - Closeout												
11.0 - Warranty Phase												
12.0 - Lessons Learned												

Level of Effort by Task Series (in hours):

TASK ACTIVITIES	A. Project Manager	B. Department Head	C. Work Scheduler	D. HR Specialist	E. Project Designer	F. Technician/CAD	G. Purchasing	H. Material Expediter	I. Fabricators	J. Accounting	K. Technical Editor	L. Clerk Typist	M. Shipping Manager	N. Sales/Marketing	O.	TOTAL HOURS
01.0 - Project Management	50	2												2		
02.0 - Quote	4	4	4		2					2	2	2		2		
03.0 - Pre-project	4	4	2	2												
04.0 - Contract negotiations	4	4												4		
05.0 - Kickoff	4	4	4	2	4	2	2	2	2					4		
06.0 - Planning/Scheduling	4	2	16													
07.0 - Design	4	4														
08.0 - Execution	4	2	4													
09.0 - Shipping/Delivery	2		4													
10.0 - Closeout	4	2	2													
11.0 - Warranty Phase	2															
12.0 - Lessons Learned	2	2	2													

LEVEL OF EFFORT PER MONTH (IN HOURS)

	1	2	3	4	5	6	7	8	9	10	11	12	TOTAL
A. Project Manager	8	8	8	8	8	8	8	8	8	8	8		88
B. Department Head		8	4	4	4	2	4		2		2		30
C. Work Scheduler		6	8	8		2	2	4	4	2	2		38
D. HR Specialist		2	2							2			6
E. Project Designer		2	8	4	4	4	4	4	2	4	2		38
F. Technician/CAD		2	2	2	4	4	4	4	2	4			28
G. Purchasing			2	4	4	4	4	2	2	4			26
H. Material Expediter		2	4	4		4	4	4	4				26
I. Fabricators		2		2	2	8	8	8	8		2		40
J. Accounting		2						4	16				22
K. Technical Editor		2								2	2		6
L. Clerk Typist		2								2			4
M. Shipping Manager							2	2	8	4			16
N. Sales/Marketing	2	4	6							6	2		20
O.													0
TOTAL HOURS	10	42	44	36	26	36	40	40	56	38	20	0	388

The information in the shaded area above is missing. Can you fill in the missing information so that all the rows & columns add up? (Be aware that there is no unique answer.)

Part 2. Your Own Project   https://app.box.com/s/gq1t096g0cs51bbi4dwzi681xutzrh72

Small Projects Tool
Schedule and Level of Effort Analysis Chart

Project:

LEVEL OF EFFORT BY TASK SERIES (IN HOURS)

TASK SERIES

TASK ACTIVITIES	MONTH 1	2	3	4	5	6	7	8	9	10	11	12	A. Project Manager	B.	C.	D.	E.	F.	G.	H.	I.	J.	K.	L.	M.	N.	O.	TOTAL HOURS
0100 - Quote																												
0200 - Pre-project																												
0300 - Contract negotiations																												
0400 - Kickoff																												
0500 - Planning/Scheduling																												
0600 - Design																												
0700 - Execution																												
0800 - Shipping/Delivery																												
0900 - Closeout																												
1000 - Warranty Phase																												
1100 - Lessons Learned																												

LEVEL OF EFFORT PER MONTH (IN HOURS)

	MONTH 1	2	3	4	5	6	7	8	9	10	11	12	A	B	C	D	E	F	G	H	I	J	K	L	M	N	O
A. Project Manager													0														
B.														0													
C.															0												
D.																0											
E.																	0										
F.																		0									
G.																			0								
H.																				0							
I.																					0						
J.																						0					
K.																							0				
L.																								0			
M.																									0		
N.																										0	
O.																											0
TOTAL HOURS																											0

Time is the scarcest resource and unless it is managed
nothing else can be managed.

—Peter Drucker

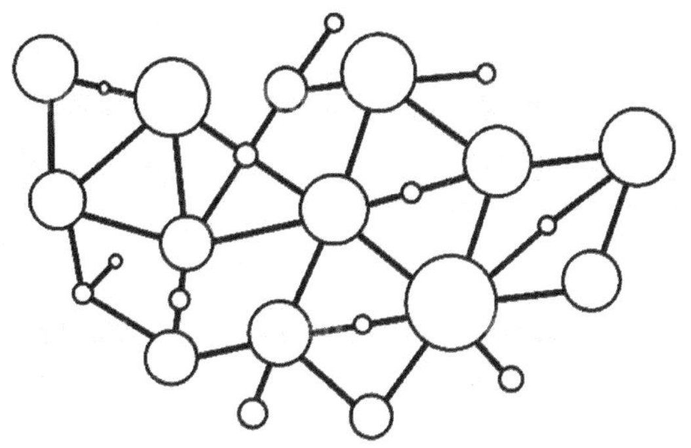

Develop a Project Network Diagram

A *project network* is a tool for planning, scheduling, and monitoring project progress. You develop the network using the information collected for the WBS.

The network diagram is like a flowchart of the project work. It depicts the project activities that must be completed, the logical sequences, the interdependencies of the activities, and, in most cases, the time for the activities to start and finish along with the longest path through the network—the critical path.

The project network is the framework for the project information system that project managers use to make decisions about time, cost, and performance.

The project network can take time to develop, and this may become costly. Some people ask, are networks worth the struggle and cost? The answer is definitely yes. The exception would be when a project is considered trivial or is short in duration.

People can easily understand a network diagram because it displays the flow and sequence of work through the project.

Once you develop a network diagram for your project, you can easily modify it when unexpected events occur. For example, if materials for an activity are delayed, you can quickly assess the impact and revise the project in only a few minutes with the use of a computer. Then you can rapidly communicate the revisions to the client/stakeholders and project team.

The project network provides added information and insights. It is the basis for scheduling labor and equipment. It enhances communication that melds all managers and groups together in meeting the time, cost, and performance objectives of the project. It provides an estimate of project duration rather than picking a project completion date or using someone's preferred completion date. The network gives the times when activities can start and finish and when they can be delayed. It provides the basis for budgeting the cash flow of the project. It identifies critical activities, which are activities that should not be delayed if the project is to be completed as planned. It highlights what activities to consider if you need to compress the project to meet a deadline.

There are other reasons that project networks can be worth their weight in gold, but in essence, having a project network minimizes surprises by getting the plan out early, and it supports corrective feedback.

A commonly heard statement from practitioners is that the project network represents three-quarters of the planning process. Perhaps this is an exaggeration, but it signals the perceived importance of the network to project managers in the field.

How to Build a Network

You build a network using nodes (boxes) and arrows. Nodes depict activities, and arrows show dependency and project flow. The activity represents one or more tasks that consume time.

The network process is similar to the WBS process. However, network diagrams illustrate the project schedule by identifying dependencies, sequencing, and timing of activities, which the WBS is not designed to do.

The primary inputs for developing a project network plan are **work packages**.

Two Approaches to Project Networks: AON and AOA

There are **two approaches** to developing project networks: **activity-on-node (AON)** and **activity-on-arrow (AOA)**

Both methods use two building blocks—the arrow and the node. Their names derive from the fact that the first uses a node to depict an activity, while the second uses an arrow to depict an activity.

From the first use of these two approaches in the late 1950s, practitioners have offered many enhancements. However, these basic approaches have withstood the test of time and still prevail with only minor variations in form.

In practice, the AON method is most often used in project planning. But there are good reasons for project managers to be proficient in both methods. For example, departments and organizations typically have their preferred approach and are loyal to software already purchased and in use. As a project manager, you should be able to use AON or AOA.

Constructing a Project Network Using the AON or Precedence Diagramming Method (PDM)

The AON method is also referred to as the **precedence diagramming method (PDM)**.

For the AON or PDM method, a project schedule network diagram uses boxes or circles, referred to as nodes, to represent activities, and it connects them with arrows that show the dependencies.

This method addresses the types of relationships between activities. These include **Finish-to-start** (FS), **Finish-to-finish** (FF), **Start-to-start** (SS), and **Start-to-finish** (SF). This identifies when one activity can start or finish in relation to another. This type of network diagramming does not employ the use of dummy activities.

In the **FS relationship** illustrated below, Activity 1 (predecessor) must finish before Activity 2 (successor) can start. In the **SF relationship** illustrated below, Activity 7 cannot start until Activity 8 is completed.

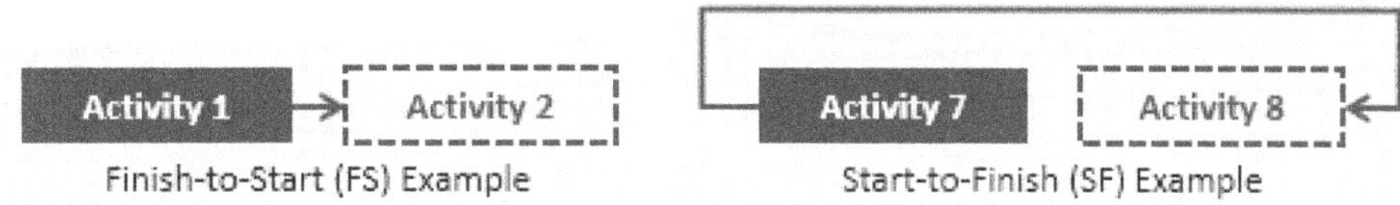

Finish-to-Start (FS) Example   Start-to-Finish (SF) Example

In the **SS relationship** illustrated below, Activity 5 must start before Activity 6 can start. As an example, you cannot drive your car until after you start it. In the **FF relationship** illustrated below, Activity 4 cannot finish before Activity 3 is finished. For example, you typically do not turn off your car until after you put it in Park or Neutral.

Another consideration of timing has to do with **lead** and **lag times**. **Lag time** is illustrated in the example below. It shows Activity 2 can start *after* Activity 1 has started.

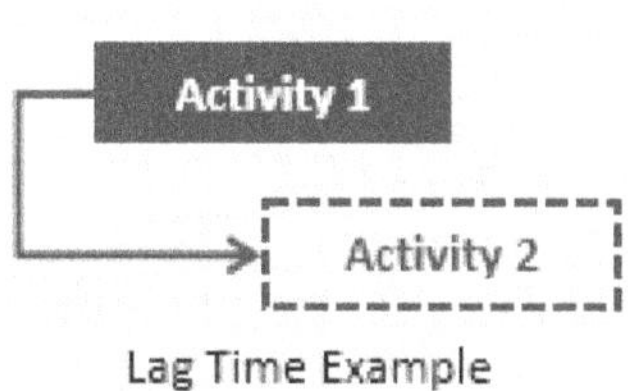

Lag Time Example

Lead time is when Activity 2 can start *before* Activity 1 has finished. As an example: Laying the foundation of a house could be Activity 1 and framing the walls could be Activity 2, whereby you could start framing the walls independently while the foundation is being built and put the framed walls in place once the foundation is completed.

In the example that follows, laying the foundation of a house could be Activity 1 and water-sealing the foundation could be Activity 2. The water sealing could start with the first part of the foundation; while the remainder of the foundation is laid, the completed portion could already be receiving water sealing.

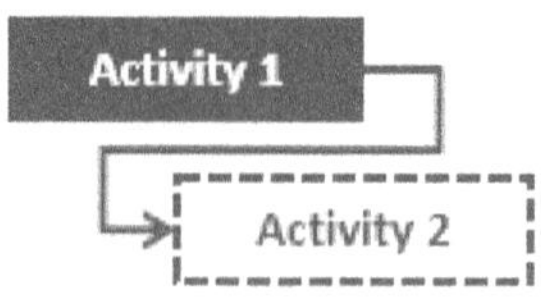

Lead Time Example

Remember, a predecessor task or activity must be completed before the next activity in the project can be done. Return to the Activity Template and review the sequence of activities. Consider the tools we have discussed: precedence diagramming, dependency determination, and lead/lag time.

Also, consider there are several types of dependencies: mandatory, discretionary, external, and internal. We need to first determine if any activity must be done before any of the others. After you list and sequence the activities, you can show the flow of activities.

Constructing a Project Network Using the AOA or Arrow Diagramming Method (ADM)

The **AOA** method is also referred to as the **arrow diagramming method (ADM)**.

In AOA or ADM networks, activities are represented by arrows. Precedence relationships between activities are represented by numbered circles connected by one or more arrows. The length of the arrow represents the duration of the activity.

Sometimes **dummy tasks or activities** are added to represent a dependency between tasks. The dummy is not an actual activity. The dummy indicates precedence that cannot be expressed using only the actual activities.

You indicate the dummy activity by using a dashed line.

A dummy task often has a completion time of 0.

The following network diagram illustrates the use of a dummy activity. Notice the dashed line.

AOA Network Diagram Example with Dummy Activity

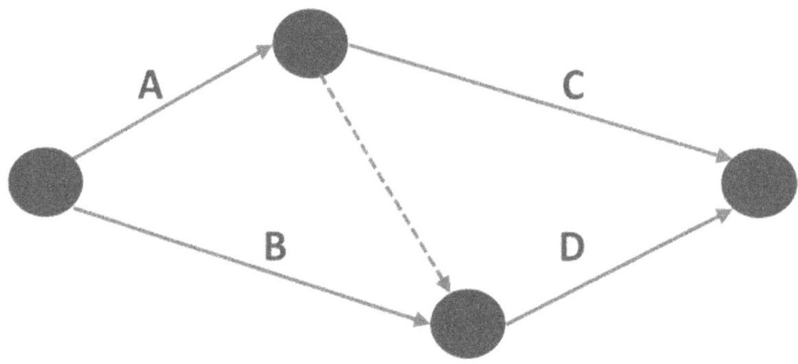

PERT and CPM

The activity diagramming method is often used for **Program Evaluation and Review Technique (PERT)** and **Critical Path Method (CPM)** analysis. However, the use of AOA or ADM as a common project management practice has declined with the adoption of computer-based scheduling tools.

The network diagram below illustrates the concept of the **critical path**. A project's critical path is the path of activities that take the longest time to complete. In this example, the critical path is the route in the middle with a total completion time of 12 weeks.

Critical Path Network Diagram Example

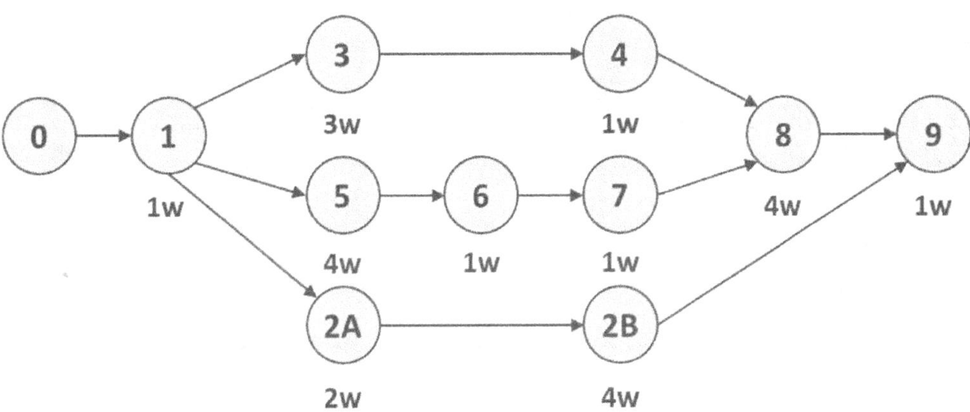

Summary of Rules for Developing a Network Diagram

The following **eight rules** apply when developing a project network:

1. Networks flow typically from left to right.
2. An activity cannot begin until all preceding connected activities have been completed.
3. Arrows on networks indicate precedence and flow. Arrows can cross over each other.
4. Each activity should have a unique identification number.
5. An activity identification number must be a larger (higher) number than any activity number that precedes it.
6. Looping is *not* allowed. In other words, recycling through a set of activities cannot occur.
7. Conditional statements are *not* allowed. In other words, do not use, "If successful, do something; if not, do nothing."
8. Experience suggests that when there are multiple starts, a common start node can be used to indicate a clear project beginning on the network. Similarly, a single project end node can be used to indicate a clear ending.

Computer Software

Be aware that all the concepts presented to this point are best applied with the use of computer software. This is also true for the statistical techniques available to a project manager in assessing project risk, such as **PERT (Program Evaluation and Review Technique)** or **probabilistic** network scheduling.

<table>
<tr><td>What to do:</td><td>Other actions:</td></tr>
<tr><td>

☐ Review the sequence of activities for a project you are involved with or have observed to validate them.

☐ Pay attention to the relationships (predecessor and successor) between activities.

☐ Create a simple network diagram for a work-related project or event.

</td><td>

☐ _______________________

☐ _______________________

☐ _______________________

☐ _______________________

☐ _______________________

</td></tr>
</table>

✓ A network diagram is a visual representation of a project schedule that uses nodes and arrows to show the relationship between activities.

✓ A network diagram shows the sequence of activities and considers the relationships between activities.

✓ Precedence diagramming is one tool you use to determine the type of relationship between two activities. There are four major types of dependencies: mandatory, discretionary, external, and internal.

✓ Developing a schedule is critical for project success. It serves as a tool for time management, and you also will use the schedule to track and measure project progress.

✓ You can create a network diagram for the overall project. If needed, you can also create a network diagram for an activity area within a project.

✓ When time is critical to your project, use a network diagram to identify the critical path. Then, working with your team, brainstorm ways to shorten the critical path.

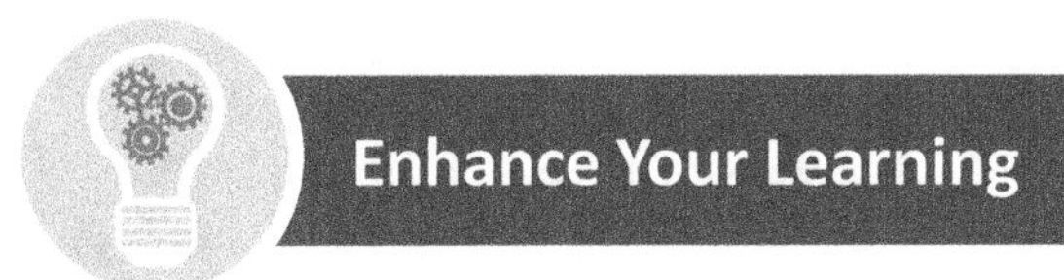

Enhance Your Learning

Watch the following 7-minute video from Creative Education to view a simple example for preparing a network diagram.

Creative Education. (2020). *How to draw a Project Network Diagram.*		Available at: https://www.youtube.com/watch?v=pWkhR67bkXw

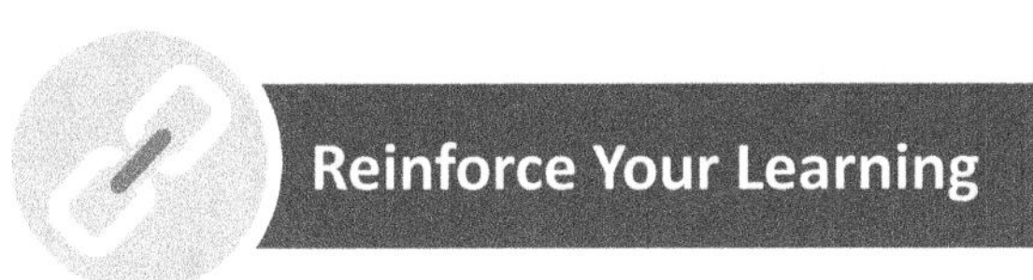

Reinforce Your Learning

Network Diagrams—Making Copies. Review the activities for the Making Copies project previously discussed. Use the activities list you made or the list below and sketch a network diagram for the project.

Activity #	Activity Description
1	Gather documents to copy.
2	Walk from desk to copier.
3	Make the copies.
4	Walk from copier to desk.
5	Distribute the copies.
6	File personal copy.

Sketch a network diagram for the Making Copies project:

Every minute you spend in planning saves 10 minutes in execution; this gives you a 1,000 percent return on energy!

—Brian Tracy

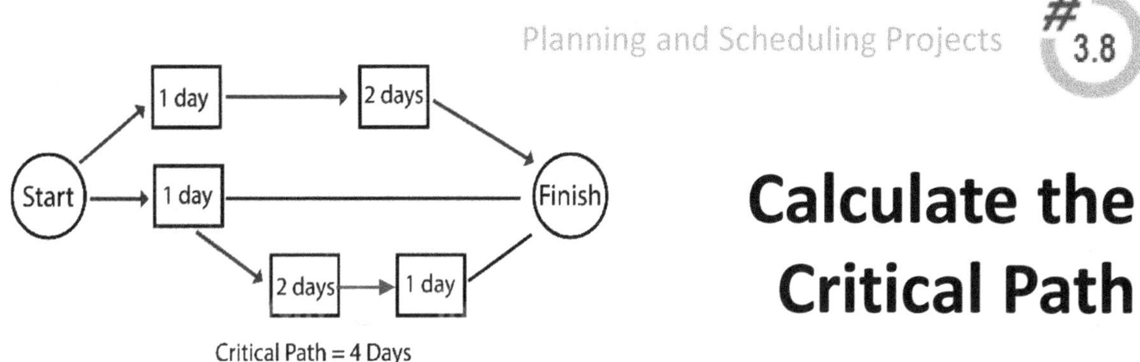

3.8

Calculate the Critical Path

Earlier in this course, we touched on how you can use a network diagram to determine a project's **critical path**. Recall that a network diagram is a visual representation of the work to be done in a project, and the critical path shows you the time (the longest path of work) needed to complete the project.

In this concept, we will go into more detail about how project managers use the critical path in project management. The critical path is of crucial concern because to manage a project effectively you must know the sequence of work activities that are the longest path (have the longest duration). It shows the time needed to complete the project.

Assume you have a project with many interconnected activities, and you want to complete the project in the least time possible because the organization has a deadline.

However, as you learned, subsequent activities cannot be completed until the work in the predecessor activities is finished. Unless you have a plan that clearly shows the relationships between the activities, you run the risk of false starts and unnecessary delays. You might have people starting before the necessary previous work is finished or sitting idle and waiting to start because they do not know the predecessor work is finished.

But even with a good plan firmly in place, you still will not know how long the project will take until you determine **the length of the critical path**. The *critical path* is the longest path through the network, and this includes both dependent and independent tasks.

The following is an example of a network diagram showing that the longest path of work is activities 0, 1, 4, 5, 6, and 7. It is the project's critical path.

Critical Path Example

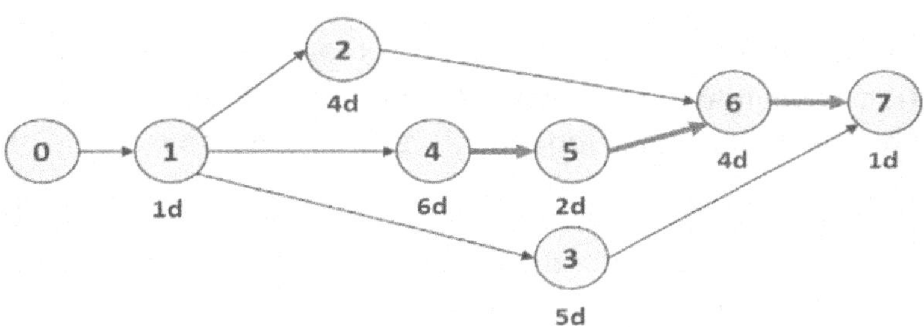

Start by noting your time estimates for each task in your network diagram. For example, you can see in the diagram above that after Activity 1 is finished, it will take four days to finish Activity 2 and six days to finish Activity 4.

Once you have your time estimates noted on your diagram, you determine the critical path by adding all the times on each of the paths. The path with the most time is the project's critical path. In this example, it is the path for Activities 0, 1, 4, 5, 6, and 7.

The Critical Path and Project Deadlines

Still using the network diagram illustrated above, the only way to shorten the length of the project is to shorten the time in the critical path (Activities 0, 1, 4, 5, 6, and 7).

Suppose the organization changes the deadline for the project by reducing the time for the project by two days. You might somehow be able to shorten the path along Activities 1, 3, and 7 by five days, but that would not shorten the project length at all. The only way to shorten the project length is by reducing its critical path.

Compressing a Schedule

Sometimes the estimated time to complete a project is longer than the time requirements imposed by the client/stakeholders. If that case, you may need to pursue other courses of action to compress the schedule.

The following is a discussion of several ways to compress a schedule. You might want to do this, for example, if you are asked to complete the project by a specific date. However, be aware that all these methods involve risk. You must ensure that the risk will not endanger the project to such an extent that problems incurred from compressing the schedule outweigh the benefits of completing the project by a specific date.

Slack or Float. *Slack* or *float* is the amount of time that some tasks in a network may be delayed without causing a delay to subsequent tasks. Identifying such tasks and removing the slack time may be an easy way to compress some aspects of a project. Related slack or float, the ***Forward pass*** is a technique to move forward through network diagram to determining project duration and finding the critical path or Free Float of the project. ***Backward pass*** represents moving backward to calculate late start or to find if there is any slack in the activity.

Crashing. This method of schedule compression uses extra resources to complete critical activities ahead of schedule. Only activities on the critical path are eligible for crashing because they affect the time needed for the project. Because crashing uses more resources, it is often more expensive and may not be compatible with the project budget.

Fast Tracking. This method of schedule compression involves rearranging the schedule so that activities that would typically be sequential are performed at the same time.

Assigning Limited Overtime. This method allows team members to work a limited amount of overtime so that critical activities can be completed earlier. The overtime must be limited to prevent team members from burn out. Too much overtime can also increase project costs.

Using Shortcuts. This method allows team members to take shortcuts rather than following exact project specifications. For example, the team might use a less expensive, generic resource instead of a name brand, or use a computer program to design a model.

Modeling Techniques

There are two modeling techniques: **what-if scenario analysis** and **simulation**.

What-if scenario analysis modeling could include the variables most likely to change and consideration of what those changes could mean to the project if they were to materialize. An example question might be what would the schedule look like if we added a person with a particular skill set? To run a model, you would plug that resource into the WBS, assign appropriate activities, and then review the overall impact it has on the project schedule.

Simulation modeling requires computer software to simulate project time, cost, or resource availability using **Monte Carlo techniques** that rely on computational algorithms and random sampling to obtain numerical results such as the probability or critically index for an activity in the schedule.

A simulation generates the **probability** of an activity or path **becoming critical**. The software uses a simple triangular distribution to represent the range and average of each activity duration.

Simulating each activity duration distribution for a project gives one set of activity values used to compute the critical path. This process is repeated hundreds of times to determine the criticality of any activity or path.

You can simulate cost by using the high and low estimates of costs for each activity and each simulation trial. By using a resource-constrained program with the simulation program, potential resource conflicts are identified and assessed.

What to do:	**Other actions:**
☐ Use the critical path to focus on activities that directly impact the project schedule.	☐ _____________________
☐ Estimates are only estimates. Monitor project progress to detect potential issues as early as possible.	☐ _____________________
☐ If time is an important aspect of the project, look for ways to shorten the critical path.	☐ _____________________
	☐ _____________________
	☐ _____________________
	☐ _____________________

Remember

- ✓ The critical path is the time needed to complete a project. In other words, you cannot complete the project sooner. The critical path is significant in every project.
- ✓ Consider your confidence level for project estimates.
- ✓ Sometimes the estimated time to complete a project is longer than the time the client/stakeholders require. If that happens, you may need to compress the schedule.

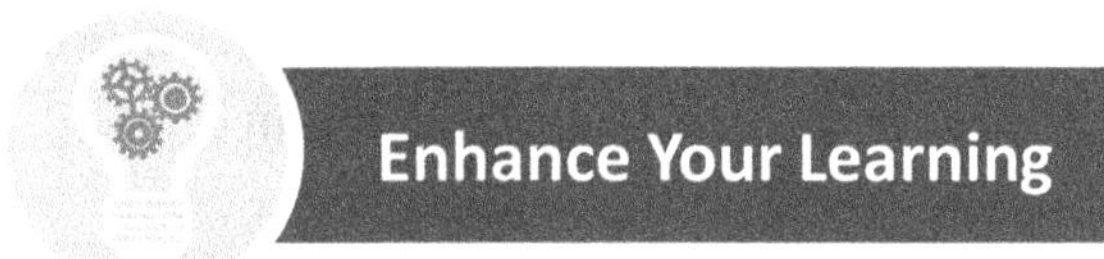

Enhance Your Learning

Watch the following 12-minute video by Andy Kaufman: How to Calculate Critical Path.

Kaufman, A. (2017). *How to Calculate Critical Path: Project Management Professional (PMP)® Exam Prep.*		Available at: https://www.youtube.com/watch?v=0Lo4zsB-bjE

Watch the following 7-minute video by Scott Christianson for a simplified explanation of project management networks and the forward and backward pass.

Christianson, S. (2019). *Project Management Networks Part 2: Forward and Backward Pass.*		Available at: https://www.youtube.com/watch?v=Momnar3Mr9Q

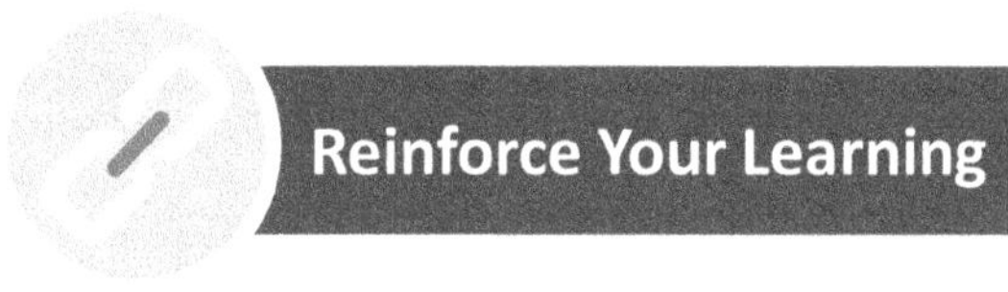

Reinforce Your Learning

Calculate Critical Path. Using the network diagram example below, calculate the number of days needed to complete the project.

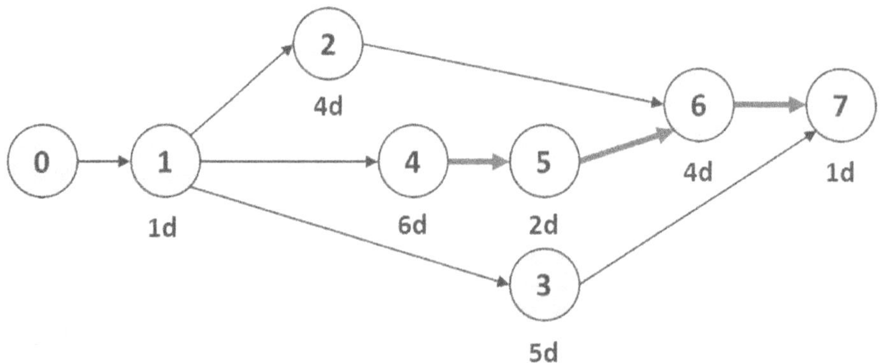

Task	Optimistic (O)	Most Likely (M)	Pessimistic (P)
Task A	2 Wks	4 Wks	5 Wks
Task B	1 Wks	2 Wks	3 Wks
Task C	2 Wks	3 Wks	4 Wks
Task B	3 Wks	5 Wks	8 Wks
Completion	8 Wks	14 Wks	20 Wks

Review the Program Evaluation Review Technique (PERT)

If you have ever developed either a project schedule or a project budget, you know it's almost impossible to come up with accurate overall estimates for either one. A simpler and better way is to estimate the individual times and costs and then add these to get your total time and cost. Doing this also helps you develop better tools for tracking your progress and monitoring your costs.

However, even for simple projects this too can be difficult. Under the best of conditions, it's often impossible to come up with a realistic schedule. Experience has shown that you can improve the accuracy of your estimate by using an approach that has proven itself in a wide range of projects. This approach takes into consideration the "best case scenario" and the "worst case scenario" and balances these two estimates with a healthy dose of the "most likely" scenario. This is a probabilistic method similar to PERT. The formula for this is shown here:

$$E = (O + 4M + P)/6$$

In this estimate approach (**E**), take your "best case scenario" or most optimistic estimate (**O**) and add it to four times the most likely estimate (**M**). Add in your "worst case scenario" or most pessimistic estimate (**P**). Divide that number by 6 to come up with the estimate. This approach is equally valid for both the schedule and the budget. Obviously for complex projects a more sophisticated approach is needed.

In 1958 the Special Office of the Navy and the Booze, Allen, and Hamilton consulting firm developed Program Evaluation Review Technique (PERT) to schedule the more than 3,300 contractors of the Polaris submarine project and to cover the uncertainty of activity time estimates.

PERT is almost identical to the critical path method (CPM) except PERT assumes each activity duration has a range that follows a statistical distribution. PERT uses three-time estimates for each activity. In other words, each activity duration can range from an optimistic time to a pessimistic time, and a weighted average can be computed for each activity.

Because project activities usually represent work and work tends to stay behind once it gets behind, the PERT developers chose an approximation of the *beta distribution* to represent activity durations. This distribution is flexible and can accommodate empirical data that do not follow a normal distribution. The activity durations can be skewed toward the high or low end of the data range.

Activity durations skewed toward the right are representative of work that tends to stay late once it is behind. The distribution for the project duration, the sum of the weighted averages of the activities on the critical path(s), is represented by a normal (symmetrical) distribution.

How to Compute the Probability of Meeting Various Project Durations

Knowing the weighted average and variances for each activity allows the project planner to compute the probability of meeting various project durations.

We describe the computation steps in the example below. The jargon may seem difficult for those not familiar with statistics, but the process is relatively easy to follow once you have worked through a few examples.

The weighted average activity time is computed using the following formula:

$$te = (a+4m+b)/6$$

where:

te = weighted average activity time
a = optimistic activity time (1 chance in 100 of completing the activity earlier under normal conditions)
b = pessimistic activity time (1 chance in 100 of completing the activity later under normal conditions)
m = most likely activity time

When three-time estimates have been specified, this equation is used to compute the weighted average duration for each activity. The average (deterministic) value is placed on the project network as in the CPM method, and the early, late, slack, and project completion times are computed as they are in the CPM method.

The variability in the activity time estimates is approximated by the equations below. The first represents the standard deviation for the activity, and the second represents the standard deviation for the project. Note the standard deviation of the activity is squared in this equation; this is also called variance. This sum includes only activities on the critical path(s) or path being reviewed.

Standard deviation for activity: $\sigma te = (b - a)/6$

Standard deviation for project: $\sigma TE = \sqrt{\sum(\sigma te)2}$

Finally, the average project duration (TE) is the sum of all the average activity times along the critical path (sum of te), and it follows a normal distribution. Knowing the average project duration and the variances of activities allows the probability of completing the project (or segment of the project) by a specific time to be computed using standard statistical tables.

The equation below is used to compute the Z value found in statistical tables (Z = number of standard deviations from the mean), which, in turn, tells the probability of completing the project in the time specified.

$$Z = (TS - TE)/ \sqrt{\sum(\sigma te)} \ 2$$

where:

TE = critical path duration
TS = scheduled project duration
Z = probability (of meeting scheduled duration) found in standard statistical tables

You can make this same type of calculation for any path or segment of a path in the network. When such probabilities are available to management, trade-off decisions can be made to accept or reduce the risk associated with a particular project duration.

At least two choices are available. First, management can spend money upfront to change conditions that will reduce the duration of one or more activities on the critical path.

A more prudent, second alternative would be to allocate money to a contingency fund and wait to see how the project progresses as it is implemented.

Program Evaluation Review Technique (PERT) Example

The following is an example of a PERT chart:

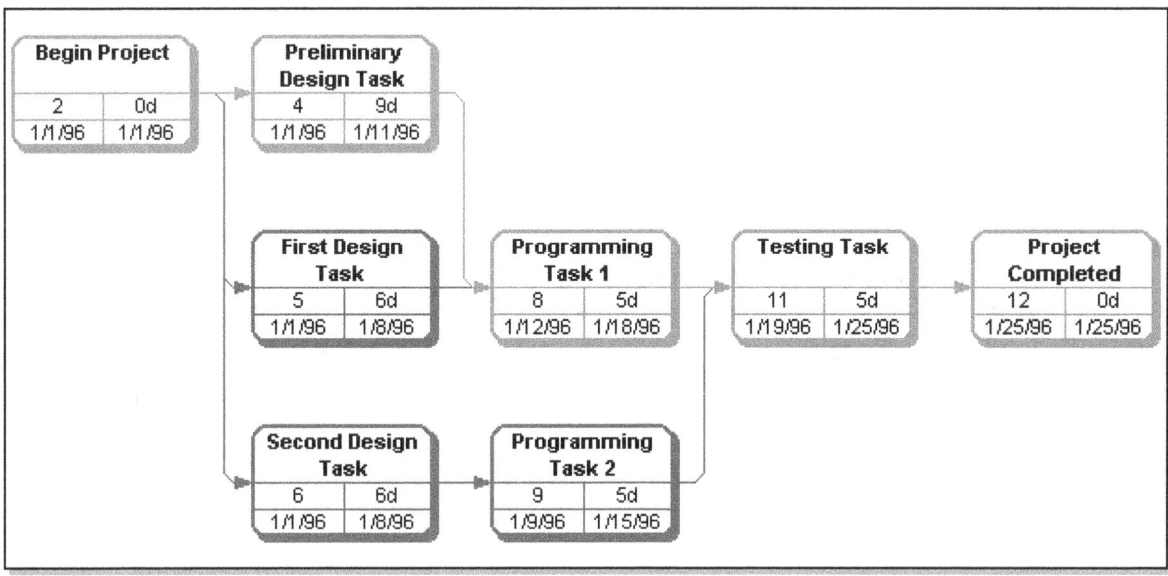

PERT Simulations

This analysis technique requires computer software to simulate project time, cost, and resource availability using the Monte Carlo technique. For example, using the same time estimates developed for PERT, simulation generates the probability of any activity or path becoming critical.

The software uses a simple triangular distribution to represent the range and average of each activity duration. Simulating each activity duration distribution for a project gives one set of activity values used to compute the critical path. This process is repeated hundreds of times to determine the criticality of any activity or path.

Cost can be simulated similarly by using the high and low estimates of costs for each activity and each simulation trial. By using a resource constrained program with the duration PERT simulation program, potential resource conflicts are identified and assessed.

Risk retention or transfer decisions are made using information gained from time, cost, and resource simulation.

PERT and PERT simulation are useful in extremely important projects that have a great deal of inherent uncertainty and reasonably accurate time estimates for activities.

Numerous software packages are available. Here are a few examples:

- Clarizen
- Microsoft Project
- Primavera SureTrak
- Project Libre
- ZOHO Projects

What to do:

- ☐ Identify a project where you are uncertain of activity time estimates.
- ☐ Review the PERT technique and consider if it may be appropriate to use.
- ☐ Ask those experienced with using project scheduling software to recommend a product.

Other actions:

- ☐ _______________________________
- ☐ _______________________________
- ☐ _______________________________
- ☐ _______________________________
- ☐ _______________________________

✓ PERT is almost identical to the critical path method (CPM) technique, except it assumes each activity duration has a range that follows a statistical distribution.
✓ PERT uses three-time estimates for each activity. Basically, this means each activity duration can range from an optimistic time to a pessimistic time, and a weighted average can be computed for each activity.

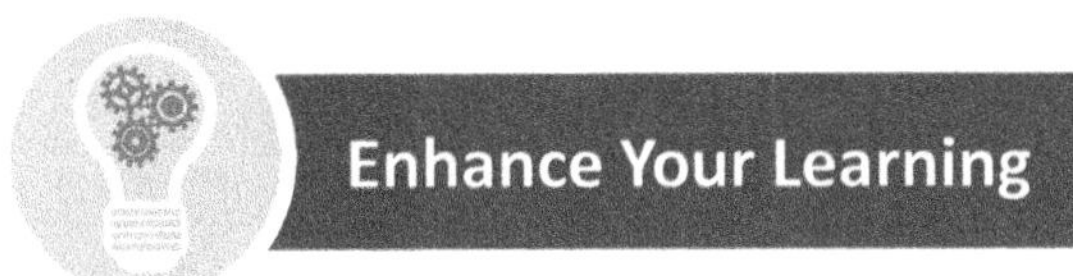

Watch the following 5-min. video by Mike Clayton to learn more about how to create a PERT Chart

Clayton, M. (2017). *What is a PERT Chart? Project Management in Under 5.*		Available at: https://www.youtube.com/watch?v=i160aaBX7mE

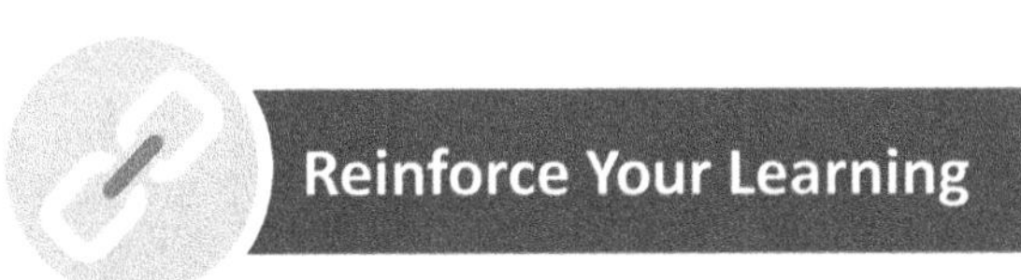

Retirement Celebration Planner. Try your hand at estimating the time it will take to plan a formal event like a retirement celebration. Use the probabilistic formula to calculate (in months) the result. Then compare your estimate with others.

$$E = (O + 4M + P)/6$$

Optimistic:

Pessimistic:

Most Likely:

Estimate =

The time to repair the roof is when the sun is shining.

—John F. Kennedy

Be Aware of Common Scheduling Pitfalls and Tips for Success

As we have discussed, one of the most challenging functions of a project manager is planning a complete and realistic project schedule. The plan lists the work to be done, by whom, and when it starts and finishes.

Without adequate planning and scheduling work, a project manager cannot manage a project because they will not know what work should be in progress or completed at any given time, and when foundational work is not done, later work cannot follow. Without an adequate schedule, the project manager may not even know this is happening and so cannot take corrective action. Without an adequate plan and schedule, the project manager cannot communicate with stakeholders and team members.

In this concept, we discuss common pitfalls to avoid, followed by 15 tips for planning and scheduling a successful project.

Common Pitfalls. Below are 15 mistakes project managers sometimes make. Any of them can cause a project to flounder and possibly fail.

1. Not adequately defining the project's purpose.
2. Not getting requirements in writing.
3. Making up or getting estimates from unreliable or unqualified sources.
4. Neglecting to reserve necessary resources.
5. Assuming needed staff members will be available and trained.
6. Failing to unite the team.
7. Not scheduling time for client and stakeholder reviews and changes.
8. Being too rigid and refusing to adapt when problems arise.
9. Underestimating project risk and assuming you can control all activities in the schedule.
10. Forgetting the roles client and stakeholder play in keeping to the schedule and the need to obtain approvals and sign-offs.
11. Assuming all preliminary groundwork needed for the next task is done.
12. Not scheduling adequate time for outside agency reviews and changes.
13. Not scheduling time for communication and coordination tasks such as phone calls, meetings, and client relationship building during and after the project.
14. Not providing adequate contingency plans in case deadlines are not met.
15. Assuming that because you are a project manager, others will respect your authority, accept your schedule, and participate accordingly.

Tips for Success

Below are 15 tips that will help you plan and schedule a project and bring it to a successful conclusion.

1. Perform due diligence to define requirements: work, people, time, and costs.
2. Ensure the right resources and subject matter experts provide estimates.
3. Create or find templates for project documents and use them.
4. Create a milestone for each project deliverable.
5. Prepare in calendar time, not in the number of workdays.
6. Build time into the schedule for slippage and revisions.
7. Involve your team in defining tasks required to produce each deliverable.
8. Create task criteria to accurately measure and track progress.
9. Break down the time needed to complete tasks and activities into 40 hours or less.
10. Assign every task to a specific team member.
11. Consider whether your team members will be assigned projects in addition to yours.
12. Ensure that the project schedule is updated as soon as possible.
13. Establish and communicate task and milestone naming conventions to avoid confusion.
14. Automate the reporting process.
15. Ask questions, and then listen.

What to do:

- ☐ Do not underestimate the importance of a complete and realistic schedule.
- ☐ Review the planning and scheduling pitfalls. Are you making any of these mistakes?
- ☐ Determine what tips for success you can apply in your work.

Other actions:

- ☐ _______________________________
- ☐ _______________________________
- ☐ _______________________________
- ☐ _______________________________

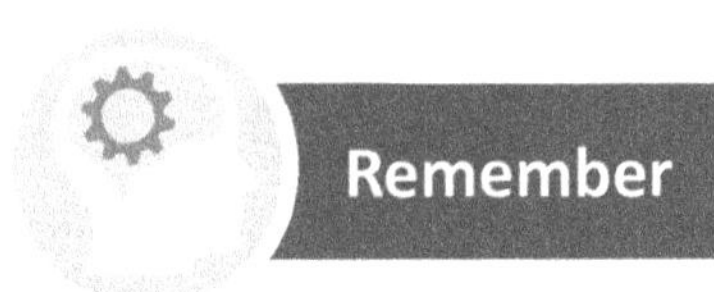

- ✓ One of the most challenging functions of a project manager is the creation of a good project schedule.
- ✓ Project managers sometimes underestimate the importance of planning and scheduling project work.
- ✓ Project managers must plan and schedule as accurately as possible and learn from past lessons and utilize the wisdom of others.

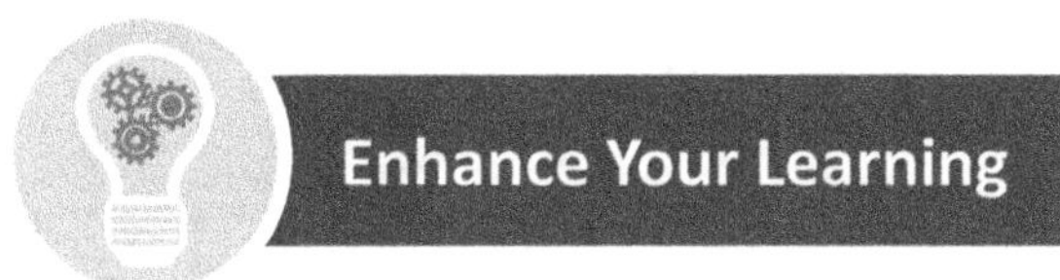

Watch the following 12-minute video by Tony Zink to obtain project management scheduling tips.

Zink, T. (2017). *Project Management Tips: Working With Tasks and Milestones in Your Project Schedule.*		Available at: https://www.youtube.com/watch?v=tP4wIlWjsqU

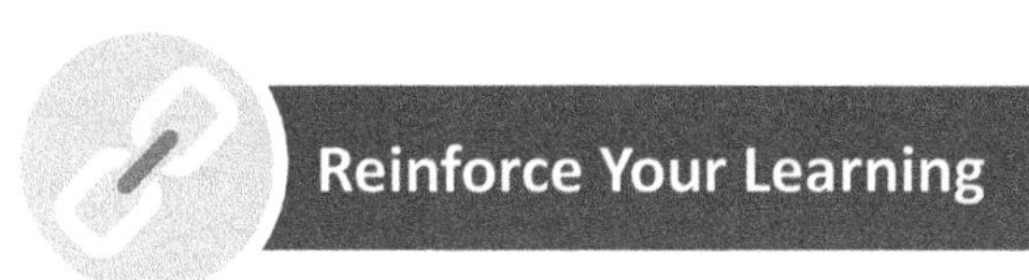

Scheduling Tips. From the list of tips, select the three you consider most important.

1.

2.

3.

204

*Unless commitment is made, there are only promises
and hopes, but no plans.*

—Peter F. Drucker

Summary
A Final Word: *Project Planning and Scheduling*

Consider what actions you can take to better understand your influence on a project schedule. All roles within a project have the potential to impact the schedule, both positively and negatively. Consider how your role may impact the project schedule and time management.

You have completed a chapter on Project Planning and Scheduling.

Information you reviewed in the chapter include:

- Describing what is included in a Schedule Management Plan
- Discussing common project scheduling tools and their applications
- Developing a project schedule
- Explaining what is meant by a schedule baseline
- Identifying common planning and scheduling pitfalls

In the next *Mastering Project Management: Planning for Performance* chapter, we will focus on Monitoring and Controlling Projects.

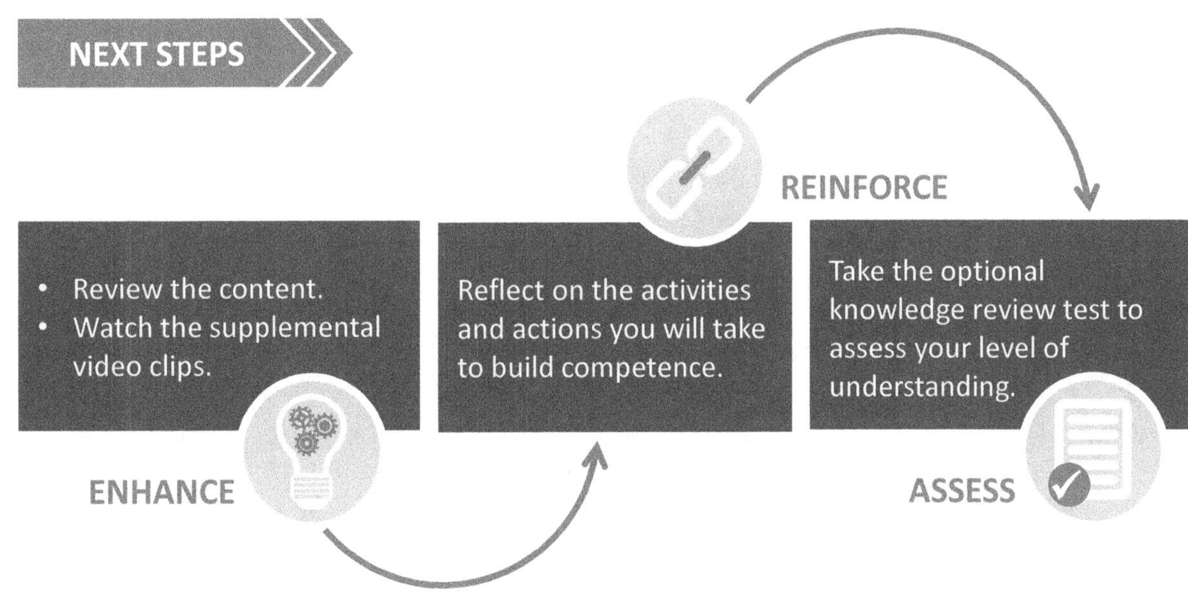

Notes:

Planning and Scheduling Projects—Recap Checklist

Half of all projects take longer than expected, so proper planning is critical. Good planning requires that you identify all activities and tasks, sequence them correctly, and accurately estimate schedules and resources.

1. Build the Project Schedule

- ☐ Identify project methods/approaches and tools you are familiar with and those to learn more about.
- ☐ Consider a baseline schedule as a formal agreement.
- ☐ Review the baseline schedule to validate that the project team can provide the deliverables as scheduled.

2. Sequence Project Tasks and Activities

- ☐ Review a sequence of project activities and validate they are accurate.
- ☐ Pay attention to how one activity relates to another (predecessor and successor).
- ☐ Inform stakeholders you reviewed the sequence of activities, and you agree or want to recommend changes.

3. Determine Effort for Tasks

- ☐ Identify your best sources of information for determining the effort required for completing activities and tasks.
- ☐ Consider using this same approach to estimate costs and develop a budget.
- ☐ Use the probabilistic or three-point estimating approach when you do not have better information.

4. Set Project Milestones

- ☐ Review a milestone chart that lists significant events or decision points with deadline dates.
- ☐ Select major events that incrementally build until your project is complete.
- ☐ Specify what and who needs to approve/decide before proceeding to the next point.

5. Develop Gantt Charts

- ☐ Select a simple work-related project and create a WBS outlining the major activities or tasks.
- ☐ Develop a Gantt or bar chart by listing the major tasks along with start dates and the length of time for each task.
- ☐ Try producing a Gantt chart using one of the many project scheduling software tools.

6. Use a Spreadsheet for Scheduling Small Projects

- ☐ List the tasks for a small project you are familiar with or have observed.
- ☐ Identify and list the individuals/functions or work units responsible for preforming the work.
- ☐ Develop a spreadsheet for your small project.

7. Develop a Project Network Diagram

- ☐ Review the sequence of activities for a project you are involved with or have observed to validate them.
- ☐ Pay particular attention to the relationships (predecessor and successor) between activities.
- ☐ Create a simple network diagram for a work-related project or event.

8. Calculate the Critical Path

- ☐ Use the critical path to focus on activities that directly impact the project schedule.
- ☐ Estimates are only estimates. Monitor project progress to detect potential issues as soon as possible.
- ☐ If time is an important aspect of the project, look for ways to shorten the critical path.

9. Review the Program Evaluation Review Technique (PERT)

- ☐ Identify a project where you are uncertain of activity time estimates.
- ☐ Review the PERT technique and consider if it may be appropriate to use.
- ☐ Ask those experienced with using project scheduling software to recommend a product.

10. Be Aware of Common Scheduling Pitfalls and Tips for Success

- ☐ Do not underestimate the importance of a complete and realistic schedule.
- ☐ Review the planning and scheduling pitfalls. Are you making any of these mistakes?
- ☐ Determine what tips for success you can apply in your work.

Action Planning | **Competency #1** >>

Planning and Evaluation

Establishes policies, guidelines, plans, and priorities; plans and coordinates with others; aligns required resources; monitors progress and evaluates outcomes; improves organizational efficiency and effectiveness.

Briefly describe how improvement in this competency will help you achieve important results or better meet your job responsibilities.

List courses, books, and independent study opportunities that could help you develop this competency.

Identify one or more people who could help you, either as a role model or source of information. Write any questions you want to ask each person

What specific steps will you take? **Start Date** **Finished**

Action Planning | **Competency #2** >>

Resource Management
Demonstrates awareness of technical resources; knows how to apply resources to achieve desired outcomes.

Briefly describe how improvement in this competency will help you achieve important results or better meet your job responsibilities.

List courses, books, and independent study opportunities that could help you develop this competency.

Identify one or more people who could help you, either as a role model or source of information. Write any questions you want to ask each person

What specific steps will you take? **Start Date** **Finished**

<table>
<tr><td>Action Planning</td><td>Competency #3 >>></td></tr>
</table>

Problem Solving

Recognizes and defines problems; analyzes relevant information; encourages alternative solutions, develops plans to solve problems.

Briefly describe how improvement in this competency will help you achieve important results or better meet your job responsibilities.

List courses, books, and independent study opportunities that could help you develop this competency.

Identify one or more people who could help you, either as a role model or source of information. Write any questions you want to ask each person

What specific steps will you take?   **Start Date**   **Finished**

214

It takes as much energy to wish as it does to plan.

—Eleanor Roosevelt

215

Knowledge Review Test

Part A. Knowledge Review Test—Questions
Part B. Knowledge Review Test—Answer Sheet

Part A. Knowledge Review Test—Questions

Planning and Scheduling Projects

1. According to a PMI study, what percentage of the projects typically completed within an organization in the past 12 months finished within their initial scheduled times?

    A. 25%
    B. 35%
    C. 50 %
    D. 65 %
    E. 75 %

2. In project management, there are typically three baselines. Which of the following is NOT one of the baselines?
    A. Schedule baseline.
    B. Cost baseline.
    C. Scope baseline.
    D. Quality baseline..

3. In project management "SME" stands for:
    A. Simultaneous Method Estimate.
    B. Subject Matter Expert.
    C. Similar Metal Extract.
    D. Simple Math Equation.

4. Project managers construct project schedules based on practical methods or "rules of thumb" which come from a variety of sources. Which of the following is NOT one of the typical sources:
    A. Project manager professional judgment.
    B. The experience of subject matter experts.
    C. Ideas from the project team and stakeholders.
    D. Friends and relatives.

5. Effort often refers to the number of labor units or man-hours needed to complete a task or activity. However, effort can also refer to other items. Which of the following is NOT one of the typical items?
    A. Work breakdown structure.
    B. Human resources.
    C. Material.
    D. Financial project needs.

6. Experience has shown that you can improve the accuracy of an estimate by using the probabilistic or three-point estimating approach. This approach has been proven in a wide range of projects. Which of the following is NOT one of the scenarios?
    A. Best-case scenario.
    B. Ideal-case scenario.
    C. Worst-case scenario.
    D. Most likely scenario.

7. The components or inputs for planning and developing a schedule typically include several kinds of information. Which of the following is NOT one of the kinds of information?
   A. Activity list and attributes.
   B. Resource list and calendar.
   C. Activity duration estimates.
   D. Actuary tables.
   E. Project scope statement.

8. A project milestone can be:

   A. A task of zero duration that shows an essential achievement in a project.

   B. Any point in the project that requires approval or a decision before proceeding to the next point.

   C. Points that represent a clear sequence of events that incrementally build until your project is complete.

   D. All the answers are correct.

9. In project management, decomposition means:

   A. The process by which dead organic substances are broken down into simpler organic or inorganic matter.

   B. Taking a deliverable, task, activity, process, or event and breaking it down into smaller, more manageable components to provide better control.

   C. The process or effect of simplifying a single chemical entity into two or more fragments.

   D. A reaction in which a compound splits up into two or more simpler substances.

10. Which of the following COULD NOT be a component of a Gantt chart?

   A. Horizontal bars.

   B. A numbered list of tasks.

   C. A timeline.

   D. Expenses.

   E. Plotted points.

11. Which of the following tools is the best to use to determine the longest time the project will take?

   A. Bar chart.

   B. Work breakdown structure.

   C. Network diagram.

   D. Project scope document.

12. Which of the following items can be tracked using a spreadsheet approach?

    A. Project tasks.

    B. Project budgets.

    C. Project timesheets.

    D. Project issues.

    E. All the answers are correct.

13. The relationship between standard deviation and risk can be best described by which of the following?

    A. Standard deviation tells you if the estimate is correct.

    B. There is no relationship.

    C. The standard deviation tells you the worst-case scenario..

    D. The standard deviation tells you the level of uncertainty.

14. An "activity" in a project network is something that requires:

    A. Planning.

    B. Time.

    C. A deliverable.

    D. A precedent.

    E. An outcome.

15. Which term refers to the least time needed to complete a project?

    A. The Network Diagram.

    B. The Work Breakdown Structure.

    C. The Critical Path.

    D. The Project Scope.

    E. None of the answers refer to the least time needed to complete the project.

16. In project management, the float of an activity is determined by:

    A. Determining the wait time between activities.

    B. Conducting a Monte Carlo analysis.

    C. Determining lead and lag.

    D. Determining the length of time an activity can be delayed without delaying the critical path of a project.

17. PERT stands for:

    A. Project Estimation Review Tactics.

    B. Program Evaluation and Retrieval Techniques.

    C. Possible Estimation of Regression Tabulation.

    D. Program Evaluation Review Technique.

    E. None of the answers are correct.

18. Dummy tasks are sometimes added to represent a dependency on what type of network diagram:

    A. PDM.

    B. ADM or AOA.

    C. IBM.

    D. LAN.

19. Which item that follows is NOT a good thing to do?

    A. Do your due diligence to define requirements: work, people, time, and costs.

    B. Ensure the right resources and subject matter experts provide estimates.

    C. Assume needed staff members will be available and trained.

    D. Create or find templates for project documents and use them.

    E. Create a milestone for each project deliverable.

20. Which item that follows is a good thing to do?

    A. Do your due diligence to define requirements: work, people, time, and costs.

    B. Create task criteria to accurately measure and track progress.

    C. Break down the time needed to complete tasks and activities into 40 hours or less

    D. Assign tasks to specific team members.

    E. All the answers are correct.

Part B. Knowledge Review Test—Answer Sheet

1. ___ A ___ B _X_ C ___ D ___ E

2. ___ A ___ B ___ C _X_ D ___ E

3. ___ A _X_ B ___ C ___ D ___ E

4. ___ A ___ B ___ C _X_ D ___ E

5. _X_ A ___ B ___ C ___ D ___ E

6. ___ A _X_ B ___ C ___ D ___ E

7. ___ A ___ B ___ C _X_ D ___ E

8. ___ A ___ B ___ C _X_ D ___ E

9. ___ A _X_ B ___ C ___ D ___ E

10. ___ A ___ B ___ C ___ D _X_ E

11. ___ A ___ B _X_ C ___ D ___ E

12. ___ A ___ B ___ C ___ D _X_ E

13. ___ A ___ B ___ C _X_ D ___ E

14. ___ A _X_ B ___ C ___ D ___ E

15. ___ A ___ B _X_ C ___ D ___ E

16. ___ A ___ B ___ C _X_ D ___ E

17. ___ A ___ B ___ C _X_ D ___ E

18. ___ A _X_ B ___ C ___ D ___ E

19. ___ A ___ B _X_ C ___ D ___ E

20. ___ A ___ B ___ C ___ D _X_ E

Chapter 4

Monitoring and Controlling Projects

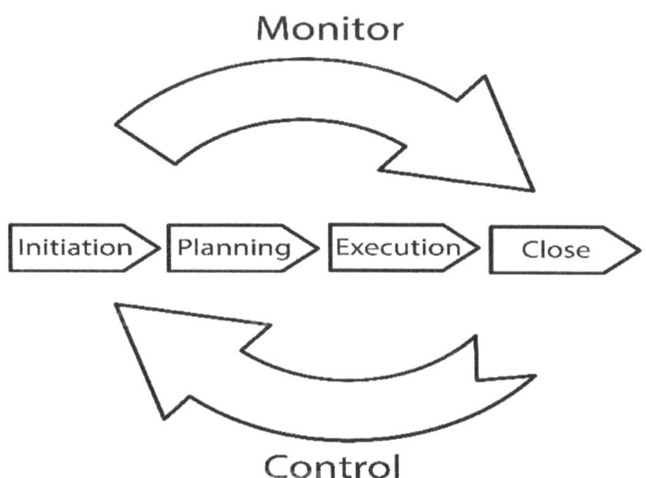

Once the project management plan is baselined, it may only be changed when a change request is generated and approved through the Perform Integrated Change Control process.

—PMBOK® Guide, 5th Edition

Overview

Monitoring and Controlling Projects

"All courses of action are risky, so prudence is not in avoiding danger (it's impossible) but calculating risk and acting decisively. Make mistakes of ambition and not mistakes of sloth. Develop the strength to do bold things, not the strength to suffer."

—Niccolo Machiavelli, The Prince

Scope changes and scope creep can undoubtedly impact a project. According to the Pulse of the Profession®: Capturing the Value of Project Management: 2015, published by the Project Management Institute (PMI), scope changes occurred on 44% of the projects. How do changes impact the project? How many aspects of the project are impacted (cost, schedule, quality, etc.)? The myriad of dynamics associated with the execution phase of a project is why monitoring and controlling is an absolute fundamental.

However, just as big of a problem can occur when project managers spend time planning the project but fall far short in monitoring and controlling the project's execution. In other words, planning is vital, but success also depends on monitoring and controlling the project work as it progresses.

Project managers must monitor and control several factors throughout the project life cycle. Developing project control baselines and gaining knowledge of tools and techniques to manage key aspects of the project are critical to its success.

In this chapter, you will learn how to:
- Manage scope variation from all project perspectives
- Identify and develop the key components of a project management plan
- Recognize the aspects of a project that most need monitoring and controlling
- Monitor and control a project
- Appreciate the benefits of early detection and correction of project metrics (scope, schedule, and cost) problems

The competencies covered in this chapter include:

Resource Management   **Management Controls**   *Oral and Written Communication*

Take Your Temperature for Monitoring and Controlling Projects

The following assessment will help you identify what you know about monitoring and controlling projects and the areas where you can improve. With you and your organization in mind, read each statement below and think about how much you agree with it. Use the numbers from 1 to 10, where **1** means you **strongly disagree** and **10** means you **strongly agree**.

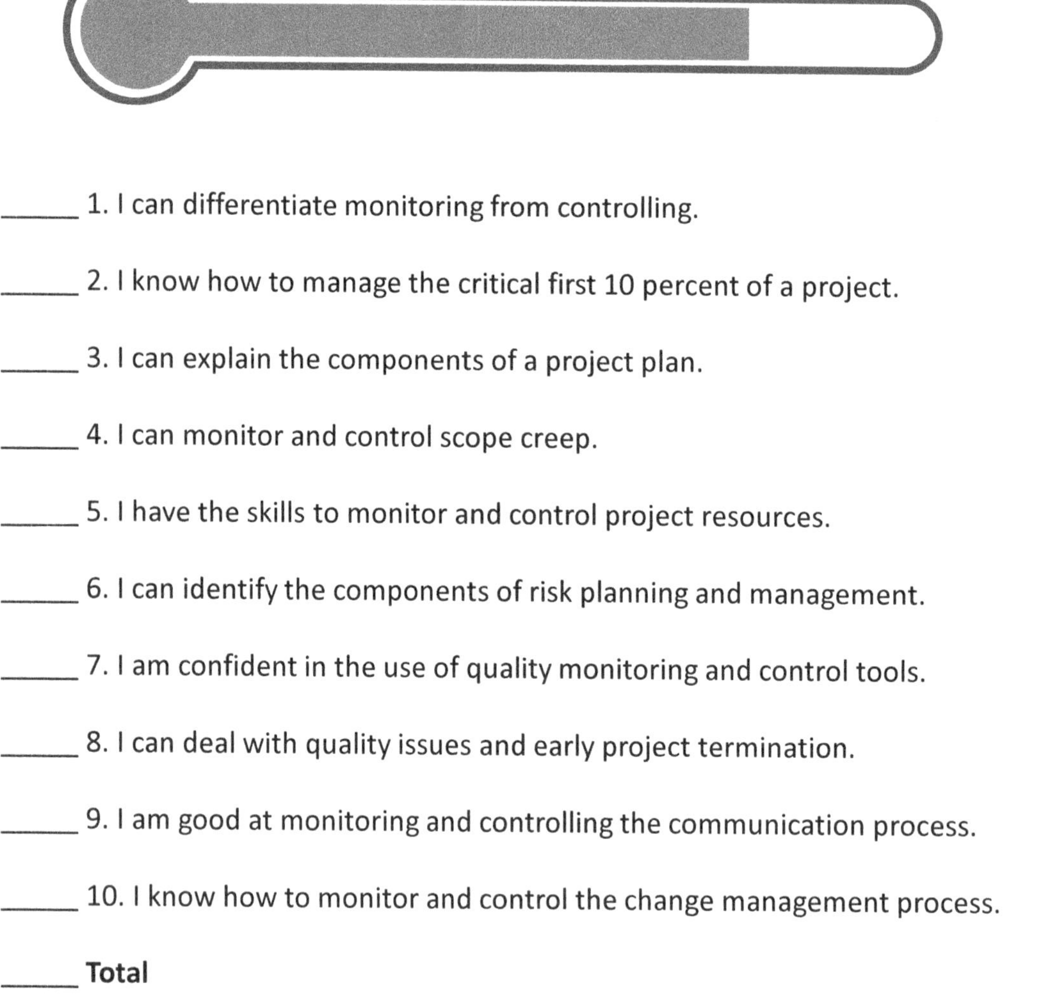

______ 1. I can differentiate monitoring from controlling.

______ 2. I know how to manage the critical first 10 percent of a project.

______ 3. I can explain the components of a project plan.

______ 4. I can monitor and control scope creep.

______ 5. I have the skills to monitor and control project resources.

______ 6. I can identify the components of risk planning and management.

______ 7. I am confident in the use of quality monitoring and control tools.

______ 8. I can deal with quality issues and early project termination.

______ 9. I am good at monitoring and controlling the communication process.

______ 10. I know how to monitor and control the change management process.

______ **Total**

What problems or situations have you experienced or observed related to monitoring and controlling projects at work?

Take a few minutes to reflect on your self-assessment. List two or three areas you want to improve.

As you progress through this chapter consider what actions you can take to demonstrate competence in the following three competency areas:

1. **Resource Management.** Demonstrates awareness of technical resources; knows how to apply resources to achieve desired outcomes.

2. **Management Controls.** Ensures the integrity of the organization's processes; promotes ethical and effective practices.

3. **Oral and Written Communication.** Makes clear and effective presentations to individuals and groups; listens to others; communicates effectively in writing; can critically review and comprehend information written by others.

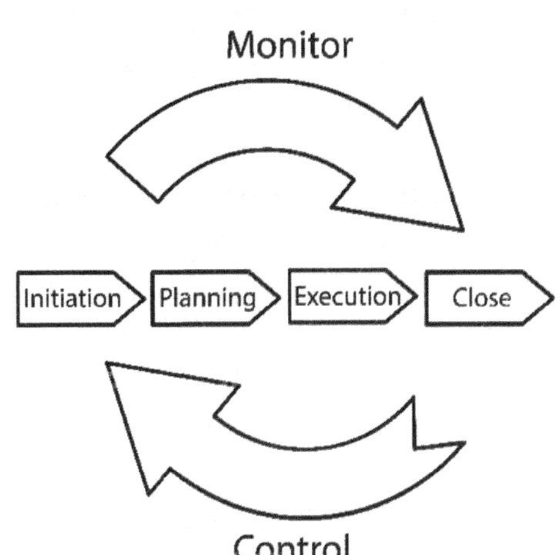

Differentiate Between Monitoring and Controlling a Project

A project will go through several phases in its life cycle: initiating, planning, executing, and closing.

This course focuses on the **execution** phase which is **doing the work**. The Project Life Cycle illustration below shows where we are in the life cycle. Execution includes both monitoring and controlling projects. **Monitoring** entails watching over the project, collecting, reporting, and disseminating project performance data and information; whereas **controlling** entails exercising influence over the project, taking action to analyze, evaluate alternatives, and recommend corrective action.

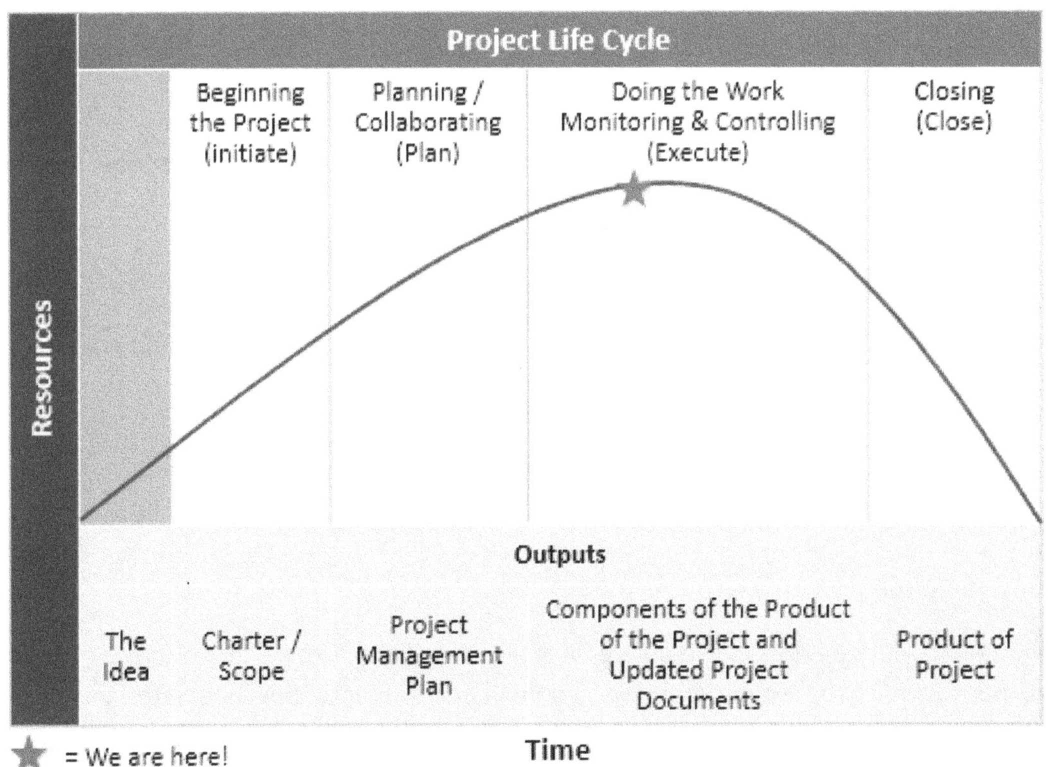

Execution is the phase where project work is done and where planning meets the realities associated with doing the work. It is also where most project funding will be spent.

In the execution phase, you follow your plan and do the work to build the project deliverables. But there is another critical job project managers must perform during the execution phase. In addition to doing the work, project managers must also monitor and control the project and make adjustments as surprises and changes occur

According to the *PMBOK® Guide*, *monitoring* involves collecting project performance data, producing performance measures, and reporting and communicating performance information. *Controlling* involves comparing actual performance with planned performance, analyzing variances, assessing trends to effect process improvements, evaluating possible alternatives, and recommending appropriate corrective action as needed.

Project Management Knowledge Areas

The execution phase involves all knowledge areas, as illustrated below. We will discuss specific tools and techniques used to determine how well a project is progressing. The intent of these tools is to facilitate early detection of anything that may have a detrimental impact on the progress. The earlier the issues are detected, the sooner the project manager and team will be in a position to address them. Some impacts could be significant and require effort to rectify.

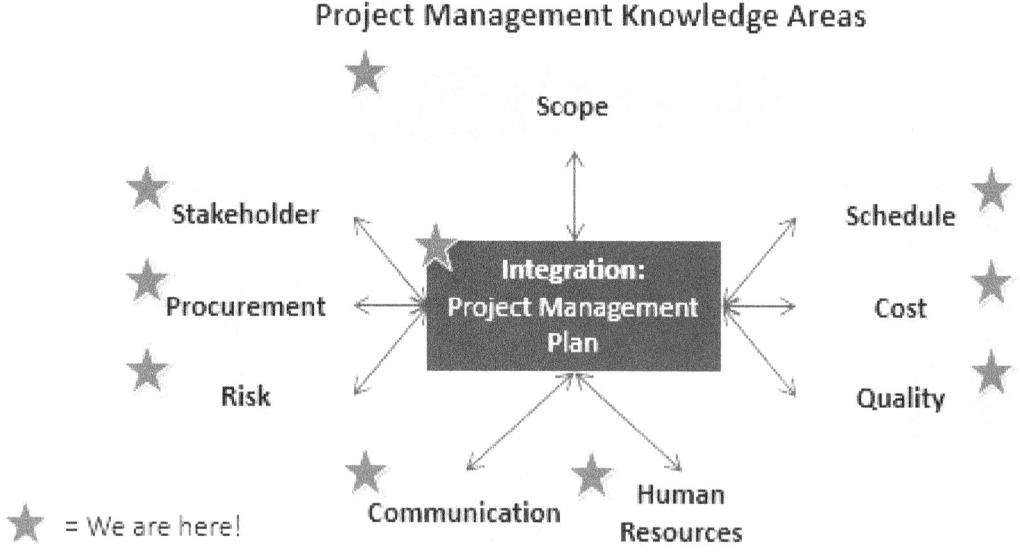

Change Management

Besides the knowledge areas in the above illustration, we will also discuss **change management**. During the execution phase, you will perform activities associated with all knowledge areas, and as the project progresses, you can expect there will be **variation** from the original **schedule baseline**, **cost baseline**, and **scope baseline**. You must ensure that all such variations are discovered and adequately accounted for.

What to do:

- ☐ Review the project charter and scope statement for updates and changes.
- ☐ Be mindful that there will be changes to the project's deliverables.
- ☐ Check the project planning for monitoring and controlling guidelines.

Other actions:

- ☐ _______________________________
- ☐ _______________________________
- ☐ _______________________________
- ☐ _______________________________

Remember

- ✓ During the execution phase, there will be changes to project deliverables. You must account for these changes in all parts of the project and in the project's documentation.
- ✓ Project documents are tools to help the project be successful.
- ✓ The project manager and team must keep project documents current, and they must accurately reflect what is actually happening with the project.
- ✓ Monitoring and controlling involve all aspects of a project.
- ✓ Every project will experience changes. Monitoring and controlling can help keep the project on track.

Enhance Your Learning

Watch the following 7-minute video by ProjectManager to learn more about controlling project scope. Then watch the 3-min. video by IPM Academy focused on the difference between project monitoring and controlling.

Whitt, J. (2013). *Controlling Project Scope: Top 5 Project Scope Control Tips.*		Available at: https://www.youtube.com/watch?v=9A80QdJPR9U

Akbar. (2022). *Project Monitoring vs Controlling, Difference B/W Project Monitoring and Controlling.*		Available at: https://www.youtube.com/watch?v=XTo99IsVS14&t=1s

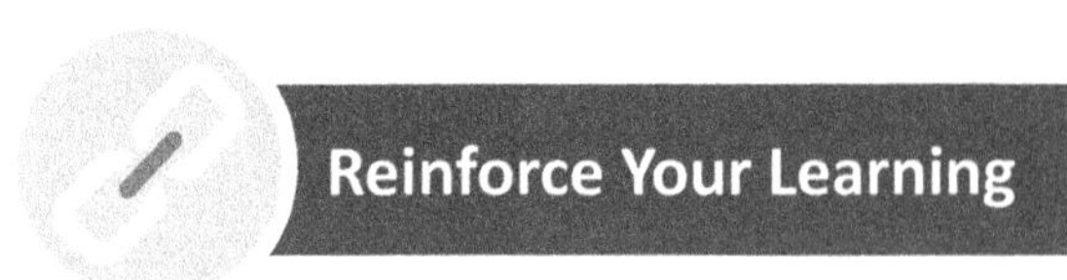

Project Changes: Making Copies. Recall our previous Making Copies exercise. What kind of surprises or changes could affect your project plan? For example, what if the copy machine is broken? What areas of the project management plan will be impacted?

Project Changes: New Home. Assume you are building a new home. After construction has started, you decide to add a two-car garage with a stone face front. You also want to advance your hobby of making pottery by having a small kiln in the back right corner. What areas of the project management plan will be impacted by these changes?

Start the Execution Phase by Reviewing the Critical First 10 Percent

In the Defining Project Scope course, we discussed the **critical first 10 percent of a project**. It is the *time spent planning a project from beginning to end.* In the execution phase, we must not only do the work planned during the first 10 percent of the project, but also ensure that the team understands the plan and regularly revisit the original baseline plan for changes.

The project plan describes what is to be done and who is expected to do it. This contributes toward the development and verification of the project's cost, schedule, and scope. Any variation from the original baseline scope is typically considered a **change request** and must be documented accordingly.

A project manager wears many hats. In addition to changes in the technical aspects of a project that need monitoring and controlling, there will also be changes in the people aspects that require monitoring and controlling. Human performance must be managed, conflicts resolved, and team members supported.

All project elements from beginning to end must be implemented, monitored, and controlled. This includes keeping all project components synchronized as changes to the scope, schedule, budget, resources, and deliverables occur.

The Execution Phase of a Project

The following list describes some of the critical jobs that must be done during a project's execution phase and associated deliverables:

Phase	Description/Purpose	Deliverables/Tools
Do Implement, Monitor, and Control Project	• Compare actual progress to planned progress • Assess and report status • Coordinate activities and resources • Implement corrective actions and update estimates and project schedule • Manage and support individual and team activities • Manage project priorities • Manage and resolve problems and conflicts • Manage changes to project scope • Assess quality of deliverables • Communicate work assignments, progress, and issues • Evaluate performance of individuals and teams	• Updated Project Management Plan Activities • Updated Risk Management Plan • Status Reports • Feedback Forms • Change Order Log • Issues Log • Interim Performance Reviews

The Critical First 10 Percent of a Project

It may seem counterintuitive, but project managers must begin the execution phase by focusing on the first 10 percent of the project, which is the time when the following planning work occurs:

- Developing the project management plan, which is also referred to as the project execution plan or simply the project plan
- Establishing design criteria
- Setting the kick-off meeting
- Defining the initial monitoring and control parameters

The design criteria and the kick-off meeting are integral parts of the Project Management Plan, but worthy of separate consideration. With comprehensive and detailed initial planning, followed by communication of the project plan and initial monitoring to ensure everything is understood and is being implemented, project success should follow, assuming the proper monitoring and control is maintained.

However, project managers sometimes neglect to monitor and control the first 10 percent of a project when they begin work on the execution phase. They mistakenly assume that everyone understands the project plan and that everyone will execute the project according to the plan. But the opposite is often true. Some team members, for example, may go off on a tangent, get ahead of themselves and the schedule, or spend too much time and detail on conceptual items. Project managers must monitor and control all project work, and this includes ensuring that the team understands the plan.

Tips to Execute a Successful Project

The following are some tips for successfully executing a project:

- Plan, and always think ahead
- Standardize
- Communicate
- Get it right the first time
- Monitor and control
- Establish a change control process

Plan, and also think ahead. Planning is critical, but at the same time, you must always think ahead and look for surprises. Did the department just get another big project that may affect your project's resources? What can be done to improve efficiency? Is someone, perhaps a team member of a stakeholder, not buying into the project? How will an upcoming holiday or the illness of a team member impact the work?

Standardize. Defined repeatable processes are one of the best ways to bring order to chaos. A simple example is using a template to create a spreadsheet as opposed to starting from scratch and having to research ideas and procedures. The project manager has more to do than to reinvent the wheel. Standardization is an important technique you can use to reduce errors and increase efficiency. Standardization also frees the project manager to think ahead.

Get it right the first time. How often does your project need rework? If you are like many project managers, you may think it is too often. Look for the reasons this is happening and ask hard questions. For example, could the need to reword parts of the project relate to the lack of clear instructions about what needs to be done?

Monitor and control. Does having to rework part of a project relate to a lack of early and ongoing monitoring before the damage is done? This often happens when a project manager thinks they are "too busy" to manage. Remember, monitoring is management.

 Change control. Does rework relate to the project manager's propensity to have it done their way even if it was not incorrect in the first place? The project manager must avoid indulging in personal whims. Nothing frustrates and demoralizes team members more than to have what they have done ripped apart and done over, and in some cases, done over again and again, just to satisfy the project manager's personal preference. Think about a recent report draft, layout, design, or drawing. How many times has it been reworked? Why was it reworked? This point is worthy of being emphasized again and again because it is so vital to project success.

What to do:

- [ ] For a project you worked on, did the team follow the project plan at the start of the execution phase?
- [ ] Revisit the tips to execute a successful project. Do you agree that monitoring is management?
- [ ] Do you often ask people to re-do work? If so, what could you do to help avoid that problem?

Other actions:

- [ ] _______________________________
- [ ] _______________________________
- [ ] _______________________________
- [ ] _______________________________
- [ ] _______________________________

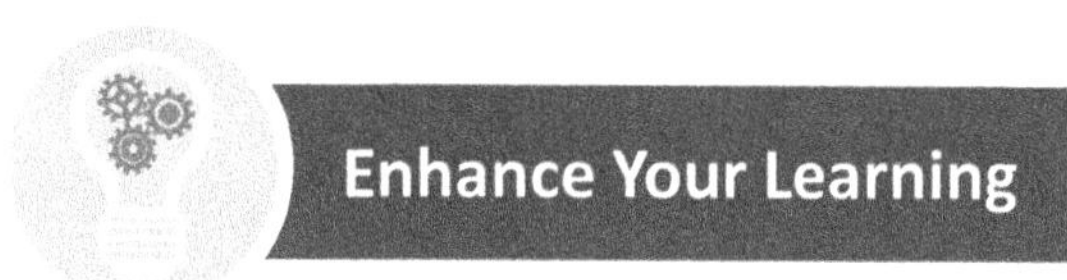

- ✓ Project integration is not easy and can be time-consuming. Make it worthwhile.
- ✓ Project integration is an iterative process. If the scope changes, other aspects of the project will also change and must be accounted for.
- ✓ The impact of project changes can ripple throughout various parts of the project.

Enhance Your Learning

Watch the following 6-minute video by Dr. Mike Clayton to learn more about project integration.

Clayton, M. (2019). *Project Management in Under 5: What is Project Integration?*		Available at: hhttps://www.youtube.com/watch?v=cHUP61KU3xE

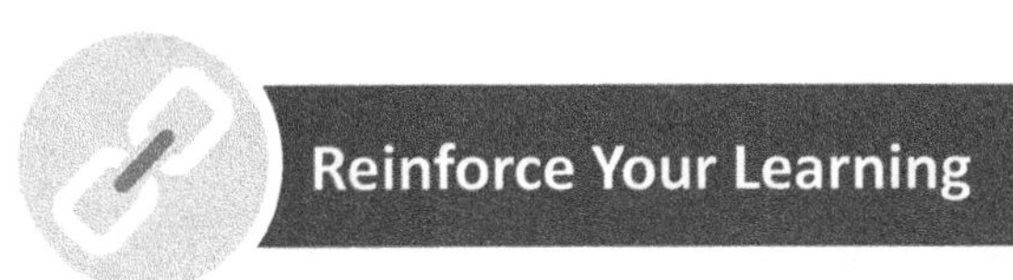

New Home: Project Management Plan Updates. Recall the New Home project from earlier. Consider how each of the following project management plan areas will be impacted by your change of adding a garage after the project had already started.

Scope:

Schedule:

Cost and Procurement:

Quality:

Human Resources:

Stakeholder Communication:

Risk:

236

Plans are of little importance, but planning is essential.

—Winston Churchill

Review the Components of a Project Plan

According to the *PMBOK® Guide, a **project management plan** or **project plan*** is a document that describes how a project will be executed, monitored, controlled, and closed.

In this concept, we will review some foundational topics having to do with the project plan: knowledge areas, changes in a project, and team performance. In later concepts, we will cover these topics in more detail as they relate to monitoring and controlling a project.

Project Management Knowledge Areas

A project management plan typically integrates all the project's component plans. PMI refers to the components as **project management knowledge areas**, as illustrated below.

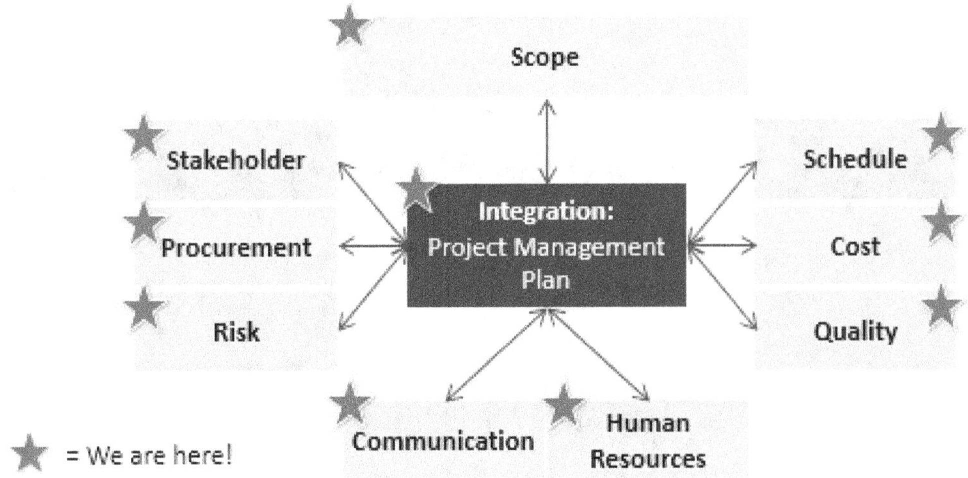

The project management plan is a structure for integrating the components and a guide for executing the project's activities. The purpose of the plan is to describe how a successful project outcome will be achieved, who will be involved in the project, and how the project will be measured and communicated. As a project manager, you must plan what you will do, how you will achieve it, and the people who will be involved.

Project management knowledge areas are described more fully in the *PMBOK® Guide.* For smaller projects, detailed plans for each knowledge area or component may or may not be needed.

Project Changes

Once a project plan is approved, the project can go forward. But as the project progresses, changes and issues will occur. Project managers must track and document them.

The following **change request log** and **issues log** examples illustrate the types of information to track and document. You will also find worksheets for these logs in Appendices A and B.

Change Request Log Example

Request Date	Change Request Reference Number	Requestor Name	Requestor Contact Info.	Request Description	Why / Benefit	Est Cost Impact	Status (APP / REJ)	Status Date	User Accept Name	User Accept Date

Issues Log Example

Issue Ref. #	Issue Description	Priority (H, M, L)	Reported by	Assigned to	Status (open / closed)	Status Date	Comments

Changes to the Original Project Plan Baseline

As issues arise and changes are proposed, the project manager must prepare a revised scope and project management plan, compare them to the original scope and plan, and communicate the differences to the client/stakeholders. This is key in controlling project change.

The purpose is to highlight the differences between the original baseline and what would become the revised baseline if the changes/issues are approved. In other words, baseline revisions show the impact of change on the original schedule, cost, and scope. Seeing the difference, the client/stakeholders will understand the potential effects the changes/issues would have on the project. They can weigh the costs of the proposed changes (in the schedule, cost, or scope) against the benefits, and make an informed decision to approve the changes or not. If needed, they can do a cost-benefit analysis to help them reach a decision.

Project Team Performance

Human performance is another crucial aspect of the project management plan. Project managers sometimes have limited opportunities in terms of deciding who will participate on a project team. Further, the project manager may not be familiar with the team's skills and any previous performance issues.

Performance reviews from previous projects can help determine if any team member issues need to be monitored and controlled in the current project. If any team members faced challenges in the past, the project manager must work to understand the problem, monitor current performance, and proactively take corrective action as needed.

The corrective action could be as simple as having a one-on-one meeting with a team member considered at risk. The project manager may learn, for example, that a person is experiencing difficulty because they are relatively new to their position or the organization, or they may be dealing with technology new to them, or they may have a conflict with someone on the team.

Once the project manager understands the source of the problem, the manager must develop remedies, such as training and coaching. Following that, the project manager must monitor performance to ensure tasks are completed as scheduled.

What to do:	**Other actions:**
☐ Document and communicate updates to your plan: why, who, what, when, and how.	☐ __________________
☐ Review your baseline project management plan and set up triggers to help monitor your project to ensure it is going as planned.	☐ __________________ ☐ __________________
☐ Control your project by proactively taking corrective measures to get back on track.	☐ __________________ ☐ __________________

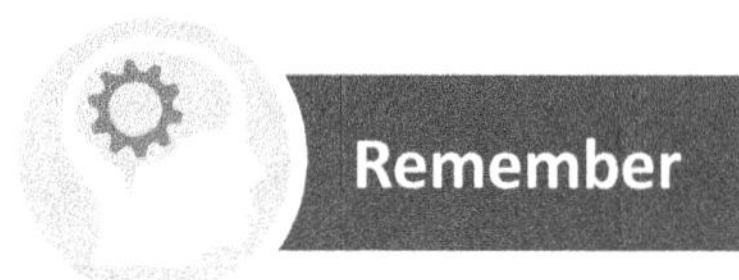

Remember

✓ Recognize that a project rarely, if ever, goes precisely according to the original plan.

✓ The published project schedule is the baseline. It is what you will be measured against, and it should be considered contractual.

✓ Change controls are the control mechanism to bridge the gap between the original baseline schedule and the actual schedule.

✓ When the schedule changes, all associated project documents must be updated.

✓ The project manager must respond quickly to issues and risks as they arise in the project and report status promptly.

✓ You can set up automated reminders or triggers to alert stakeholders of project changes.

✓ Performance shortfalls or problems in the team must be addressed as soon as possible after they occur.

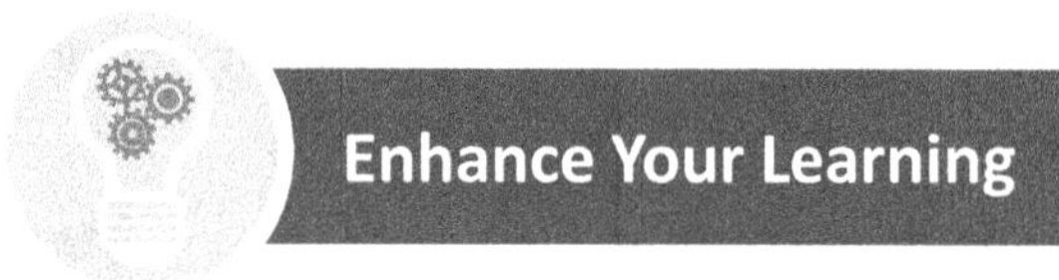

Enhance Your Learning

Watch the following 6-minute video by ProjectManager to learn more about how to successfully execute a project plan.

Bridges, J. (2019). *How to Successfully Execute a Plan.*		Available at: https://www.youtube.com/watch?v =802yQd8TNf8&t=1s

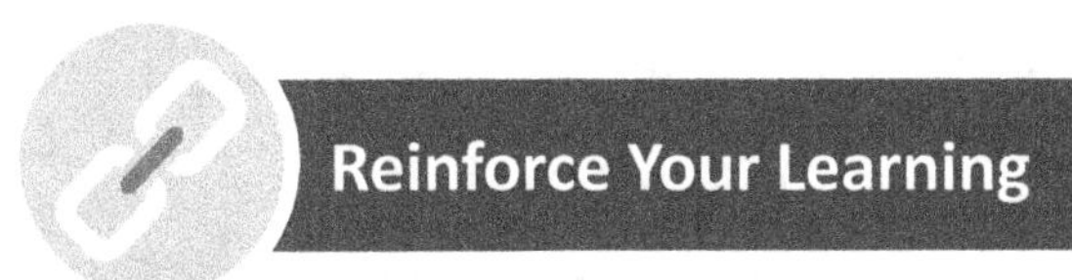

Baseline Comparisons: Making Copies. Below is the original budget (schedule and costs baseline) for our previously discussed Making Copies project.

Budget (baseline):

Activity	Estimated Duration (Sec.)	Comment	Resource
Gather hard documents to copy	35		You
Walk from desk to copier	105		You
Make 10 copies	71		You
Walk from copier to desk	90		You
Distribute the hard copies	475		You
File personal hard copy	53		You
Total Projected Schedule	829		
Minutes	13.82		
BUD Summary:			
Labor:	5.54		
Material:	0.10		
Equipment:	0.05		
Total Projected Cost:	$ 5.69		

The illustration below shows detail for the labor, material, and equipment costs.

Supporting Calculations:

40	Hours per weekend worked	**Labor**		
52	Weeks per year paid		13.82	Minutes of Labor
2,080	Hours per year paid		0.40	Per minute labor rate
50,000	Annual Salary		5.54	Labor costs
24.04	Rate per hour			
0.40	Rate per minute	**Material**		
			0.01	Cost per piece of paper
			10	Number of copies
			0.10	Cost of material
		Equipment		
			0.005	Copy / click cost
			10	Number of copies
			0.05	Cost of equipment

Project Changes. Some budget variations occurred during the execution of the copying activities.

First, on the way to the copier, you were stopped in the hall about an unrelated issue, which added to the walk time to the copier. Second, while you were at the copier, it ran out of paper, and then the new paper got jammed.

Actual:			
Activity	**Estimated Duration (Sec.)**	**Comment**	**Resource**
Gather hard documents to copy	35		Me
Walk from desk to copier	105		Me
Caught in the hallway	300	(5 minutes)	Us
Make 10 copies	71		Me
Out of paper	60		Me
Paper jam (5 pages lost)	120	(2 minutes)	Me
Copier guru helped	120	(2 minutes)	Joe
Walk from copier to desk	90		Me
Distribute the hard copies	475		Me
File personal hard copy	53		Me
Total Projected Schedule	1,429		
Minutes	23.82		
ACT Summary:			
Labor:	9.54		
Material:	0.15		
Equipment:	0.08		
Total Projected Cost:	$ 9.77		

Actual activities

So, the result was that you spent additional time because of loading paper and clearing the paper jam.

To clear the paper, you received help from a resource not included in the baseline. Time for this person (we assume the same salary as yours) was added.

Also, neither the paper used nor the additional cycles (clicks) on the copier were in the baseline, so we added them to the actual schedule and costs.

The illustration that follows shows the detail for how the help received and additional clicks were calculated.

Actual: Supporting Calculations:				
40	Hours per weekend worked	**Labor**		
52	Weeks per year paid		23.82	Minutes of Labor
2,080	Hours per year paid		0.40	Per minute labor rate
50,000	Annual Salary		9.54	Labor costs
24.04	Rate per hour			
0.40	Rate per minute	**Material**		
			0.01	Cost per piece of paper
			15	Number of copies
			0.10	Cost of material
		Equipment		
			0.005	Copy / click cost
			15	Number of copies
			0.075	Cost of equipment

Actual Supporting Calculations

The data for the calculations have been provided, so the impact of the variations on the performance results is visible. In the world of project management, this would be considered a failed project. The project failed on two of the three constraints: The schedule was elongated, and the costs were over budget.

As you can see, the impacts are significant, considering the minimal timeframe in the example. More complex issues have the same types of considerations, and the significance of the increased time/schedule and costs have a higher impact proportionally.

Discussion

This example is for a small project. There was no time to monitor and control the Making Copies project, but the concepts of having a baseline and actual data collection along the way still apply. With a project schedule in days instead of in seconds, you can see how the monitoring and control would be accomplished if similar unanticipated activities were to occur. Corrective actions could then be activated.

Activities and attributes (exhaustive list, resources, durations, relationships, and costs) **are the core foundation of the schedule.** Anything that impacts any one of these could influence the delivery of the project's product, service, or result.

Humans are eternal optimists; we generally assume a best-case scenario. For life in general, this may work well; for projects, not so much.

When estimating, you must consider any risks that could challenge the duration you are projecting. You must also be comfortable with what you are being asked to schedule.

Anything new represents a risk to a project. A new technology, a new project team, a new external entity—anything unfamiliar adds risk. Be sure to consider all the "new" in your confidence factor and provide adequate reserves (buffers) for each.

The world of projects is subject to the domino effect. When one project shares a resource with a second project, and the first project starts to go off schedule, it typically will have a detrimental impact on the second project. Now the resource is under pressure to perform more (spend more time/effort) on the first project without slowing down the second project. This situation often impacts performance on both projects.

Activity:
Reflect on this Making Copies project example, then consider a project that you are familiar with that experienced variations from the plan. Briefly describe below what was done or should have been done and list your conclusions below:

Monitor and Control Scope Creep

Recall, the **project scope** describes what is included in the project and what is not included. The **scope identifies the deliverable(s)**, which could be a product, service, or result. It will not exist until the project is completed.

The scope drives the activities list, risks, resources, costs, and schedule. Because of the potential impact of a scope change to the project plan, as the project progresses, the project manager must monitor, control, and verify all activities for variation.

The **scope** also dictates **what must be done** to produce the project's product, service, or result. As scope variations occur, they are bound to impact the schedule, costs, risks, human resources, and potentially all aspects of the project baseline (original schedule and resources).

It is imperative that the client/stakeholders clearly understand the scope of the project because they are the people who must review and approve changes to the scope. As we have discussed, once any changes are reviewed and authorized, the project can continue, and the project manager will then monitor and control project work as per a revised plan.

Uncontrolled Project Changes

PMI's The *Pulse of the Profession®: Capturing the Value of Project Management: 2015* included survey results for this question: In your estimation, what **percentage of the projects completed** within your organization in the past 12 months **experienced scope creep** or **uncontrolled changes** to the project's scope? The answer was **44 percent**! You can read more about the *Pulse of the Profession* at *www.pmi.org/pulse*

Uncontrolled changes that impact the scope can occur at any point after a project begins. Uncontrolled changes are often referred to as **scope creep**, **scope leap**, or **gold plating**. All these terms refer to scope changes after the original scope baseline has been established. They often occur when the scope of a project is not adequately defined, documented, and controlled.

Gold plating means intentionally adding extra features or functions to the project that were not included in the original scope statement. Gold plating is common and typically done by team members to show their abilities, or by the project manager to please the client/stakeholders.

How to Manage Changes in Project Baselines

The **scope baseline** is the **original scope** that subsequent project plans use as input. As part of the development of a project communication management plan, the scope baseline is used to determine who needs to be included in project communications. **If the scope changes** via a scope change control, the **communication plan** should be updated accordingly. Additionally, **corrective action** should be implemented to bring future project performance into line with the project plan.

For reference purposes, we will identify the original scope baseline as scope baseline 1. Then as changes occur, we will refer to them as scope baseline 2, scope baseline 3, and so on.

What determines whether a new scope baseline must be established depends on the organization and the project. Guidelines can detail how scope baseline levels progress from one to another. For example, a guideline could require a new baseline if the scope is changed by x additional items added, the costs associated increase by x %, or the schedule is padded by x days. This can be taken a step further by requiring an additional authorization level when any such threshold is reached.

Review the illustration below. It begins by listing the **project's original scope**, which is **scope baseline** 1. Scope changes cause a 5 percent increase in project cost. Now, per the thresholds, we must have a VP-level signature to authorize the scope changes and subsequent increase in cost. With that signature, we are now at **scope baseline 2**. The threshold and additional signature requirement call attention to what is deemed critical. In this example, the cost reaching a threshold was the trigger for an additional review of the project.

Scope Baseline Thresholds			
Scope Baseline Reference Number	Cost Increase %	Schedule Increase %	Required Approval Signatures
1	Original	Original	Sponsor
2	≤5	≤2	VP
3	>5 and ≤10	>2 and ≤5	Senior VP
4	>10	>5	President

Again, the particular thresholds and threshold levels will vary from organization to organization, and possibly from one project to another within an organization. However, the concept of scope monitoring and controlling is consistent from the standpoint of establishing thresholds and having mechanisms in place to enforce them. You should set thresholds at levels where timely corrective action can be taken.

Scope Management Plan and the Scope Verification Process

For large complex projects, the **project management plan** should include a **scope management plan**. It outlines how the original scope was determined, how it can be changed, and how scope monitoring and controlling activities will be conducted.

The **scope verification process** provides a mechanism for establishing and documenting the level and extent of completion should unexpected events occur. One example would be the termination of a project due to loss of funding.

Scope verification also involves reviewing work products and results to ensure that all were completed satisfactorily and formally accepted. A common way to validate scope changes is to do an **inspection**.

Requirements Traceability Matrix

For some projects, the scope may change because the project **requirements** change or evolve. This is often the case with software projects or the innovative design of new products and services. Some circumstances may merit the **creation of prototypes** to establish requirements.

In these situations, it is beneficial to develop a **requirements traceability matrix** for use in the monitoring and control efforts. It will help you ensure that all requirements have been satisfied.

The matrix also helps to provide verification of the project deliverables. The matrix documents signoff of the requirements associated with the deliverables (if they are not the deliverables themselves).

The following is an example of a requirements traceability matrix. This is just one format for this type of matrix. The exact information to include depends on the needs of the project and the organization.

The format may also vary according to the complexity of the project.

Requirements Traceability Matrix Example

Requirements Traceability Matrix						
Project Name:				Company:		
Project Manager:				Department		
Project Sponsor:						
Requirement Identification Number	Requirement Description	Requirement Document Reference	Test Case Reference	User Signoff	Signoff Date	Comments

Monitoring and Controlling Change

Change requests and unexpected **project issues** are a natural part of the process for monitoring and controlling scope. The project management plan and the change management plan address how changes are to be managed. This includes what type of changes are acceptable and who has the authority to accept or reject proposed changes. This may include authorization levels, depending on how the cost and schedule are impacted by the change. All this activity should be captured in the change request log and the issues log.

The **change request log**, which we described earlier, should include all change requests, whether eventually approved or not. This log can be used to determine how much change request activity is currently going on to project how much more change activity will be needed. Project issues may or may not rise to the level of making formal changes. As we have discussed, the consequences of the changes (additional scope, cost, and schedule) must be clearly understood by those approving the changes.

The people who **approve changes** must understand there are tradeoffs between allowing the changes and the impact of the changes on scope, cost, and schedule. If a higher-than-anticipated number of change requests keep coming, it may be time to add reserves for the activity and re-baseline the project accordingly.

Applicable **work performance data** should be updated according to the results of inspection and validation activities. This should include both human performance and enterprise resources involved with the project.

Updated project documents include any portion of the project that is impacted by the scope monitoring and controlling activities. This could include any documents associated with integration, scope, schedule, cost, resources, quality, communications, risk, procurement, or stakeholders.

What to do:

- ☐ Review the requirements for your project and validate them to this point in the project.
- ☐ Review the project change request log to ensure appropriate actions have been taken.
- ☐ Review the scope baseline thresholds to ensure compliance. Create thresholds if they do not exist.

Other actions:

- ☐ ______________________________
- ☐ ______________________________
- ☐ ______________________________
- ☐ ______________________________
- ☐ ______________________________

Remember

✓ Requirements are one of the best resources for monitoring and controlling a project's scope.

✓ Inspection is an effective way to validate scope.

✓ Scope baseline thresholds are a way to establish warnings.

Enhance Your Learning

View the following 8-minute video by RiskDoctorVideo about scope creep: *The Wasa–a true story of scope creep.*

RiskDoctorVideo. (2015). *The Wasa - a true story of scope creep.*		Available at: https://www.youtube.com/watch?v=kmJ59yyYza4

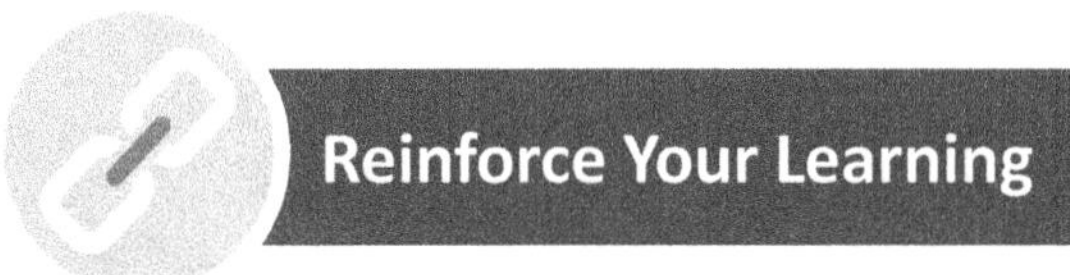

Reinforce Your Learning

Scope Creep. Consider a project where you experienced or observed scope creep. With the project in mind, respond to the questions below.

Project name:

Did the scope creep affect the cost of the project?

Did the time to do any tasks change? If so, did the project schedule change?

Did the scope creep affect the quality of the project's product, service, or result?

Nothing is less productive than to make more efficient what should not be done at all.

—Peter Drucker

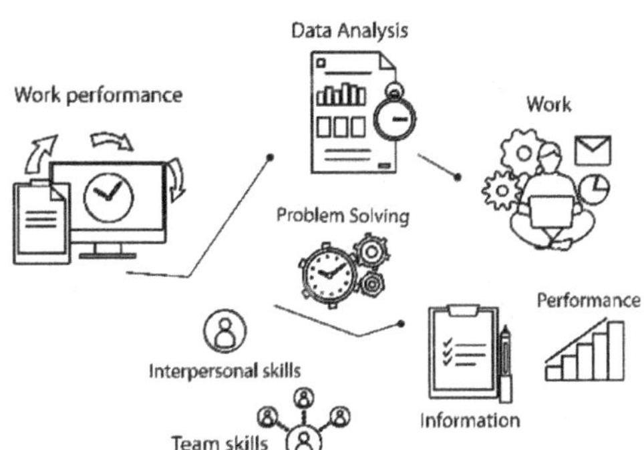

Monitor and Control Project Resources

Everything in business is governed by money. Even the most amazing project must stay within budget, be cost-effective, and justify expenditures. Knowing how to plan, track, and explain project costs are necessary skills for project managers because every project manager, at some point, will need to develop, approve, and communicate project costs and budgets.

The project sponsor/client must approve a project budget before the project starts, and then, as the project progresses, the project manager must keep the project within its budget. If a budget issue develops, the project manager must understand the cause and explain the problem.

We will go into more detail about project costs and budgeting in a later course. In this concept, we will discuss monitoring and controlling project resources because resources are significant costs in any project.

Resources must be available on time for a project to be completed successfully. Thus, planning for when resources will be needed is a crucial aspect of running a successful project.

Once the initial planning is done and the project progresses, the project manager must ensure that resources are allocated as planned.

Monitoring activities provide insight into when resource availability may be in question. Controlling activities facilitate the corrective actions needed to deal with problems.

We will examine **resources** from two perspectives: **human** (people) and **enterprise** (other than people).

Human Resource Availability

The **human resource management plan** is part of the project management plan, or at least has some elements within it. It addresses how human resource monitoring and controlling will be done.

Project resource assignments and **resource calendars** can be referenced and shared among projects.

Resource monitoring ensures that the intended allocation of resources occurs and that the project stays on schedule. For example, if a person with a specialized skill such as a nurse is allocated 50 percent to your project, but is only working 25 percent of that time, your project may fall behind schedule because the tasks will take twice as long to complete. Further, if the tasks are predecessors, the successor tasks will also push to later dates.

Besides affecting the schedule, a problem like this could affect the resource calendar, costs, and scope. The resource calendar could change not only because of changes in specific task assignments but also because of the associated predecessor or successor tasks, and that would impact future resource allocations. Costs may change because of the delay in tasks being completed. In this example, the resource allocation change would not change the scope.

Alternatives to solving the problem of a change in a resource allocation should be reviewed, and the project manager should measure the trade-offs for each. Whatever the decisions, the updated project documents need to capture the ramifications of the change.

Enterprise Resource Availability

Enterprise resources must also be available on time for the project to be successful. For example, if a specialized piece of equipment is needed to run samples for a project, and the samples are the predecessor for a lengthy analysis process, then delays in the samples will delay the analysis and potentially the entire project schedule.

By monitoring and controlling such equipment resources, a project manager would likely discover the problem and have time to find a solution before the project schedule was affected.

How to Monitor and Control Project Resources

Project performance reports track how a project is performing against its intended schedule, scope, and cost baselines. For example, if a resource spends 50 percent of their time on the project as allocated, but all their tasks take longer than anticipated, it would be prudent to re-estimate the remaining tasks. This would be considered a re-baseline of the schedule because of longer times than initially planned to complete tasks.

Observation and discussion can be effective ways of determining how resources are being used. This can include interactions in project meetings, but it may also include a less formal approach of chatting with the project team and stakeholders about how things are going, both the good and the not-so-good. An informal interaction provides an opportunity for someone to discuss an issue they feel uncomfortable talking about in front of a group.

Conflict management is a necessary part of all projects. For example, problems will occur if various team members do not agree on how to accomplish a task or do not agree with the project plan. A related issue is that a team member could do something in a way that causes others to spend more time than projected, which could lead to a problem in the schedule.

As a project manager, you must identify the source of conflict and find a solution that not only keeps the project on schedule, but that is agreeable to the people involved. Brainstorming is an excellent way to moderate conflict and encourage the team to reach a consensus on the best solution to an issue.

Conflict can occur between two individuals, such as when one person does not agree with another, or among several individuals. The symptoms of conflict can take many forms; however, **poor communication** is often the root cause.

Conflict often involves content, process, or interpersonal issues. **Content conflicts** are based on the **subject matter** of team discussions. For example, people on the team may have differences about an idea discussed in a team meeting. **Process conflicts** occur when people disagree about **how** the team should accomplish a task. For example, some people may want to do a job a certain way while others do not. **Interpersonal conflicts** are based on **relationships** between people. For example, one team member may dislike another person's communication style.

Research shows that people approach conflict in various ways. Kenneth Thomas and Ralph Kilmann, well-known for their work in conflict resolution, have observed that people tend to respond to conflict with **assertiveness**, where they strive to satisfy their own concerns, or they respond with **cooperativeness**, where they endeavor to satisfy the concerns of others.

People also tend toward particular styles of assertiveness or cooperativeness. The five basic styles, according to Thomas and Kilmann, are:

- **Competing** (assertive and uncooperative) People using this style take a firm stance on getting their needs met, and often draw on power, persuasion, or rank.
- **Accommodating** (unassertive and cooperative) People using this style sacrifice their own concerns to satisfy the concerns of others.
- **Compromising** (between competing and accommodating) People using this style look for the middle ground. They address issues more directly than with *avoiding,* but not to the extent as with *collaborating*. Concessions are often called for.
- **Avoiding** (unassertive and uncooperative) People using this style defer by circumventing, postponing, or withdrawing from tense situations. Nothing is gained from this approach.
- **Collaborating** (assertive and cooperative) People using this style try to bring all parties together to satisfy all concerns. These people are leaders who look at what each person needs and has to offer and attempt to end with a win-win for all. This problem-solving technique is the best way to resolve a conflict, but it is not always easy, especially when emotions run high.

To address conflicts more consistently across projects, organizations sometimes develop conflict management policies and procedures. This helps ensure that disputes are resolved productively. Different types of conflict may require different approaches.

Interpersonal skills are essential for all team members. The better the team can communicate within itself, the more effective the entire team will be. The project manager also must communicate effectively with stakeholders.

Project team members do not always report directly to the project manager. When they do not, the project manager must rely on persuasiveness and be creative in engaging the team and in helping them make the project a priority.

Coaching and training can significantly impact the success of a project. While traditional training can increase productivity by over 20 percent, training and coaching combined can result in nearly a **90 percent increase in productivity** (as cited in TCS, 2012). When you know how to coach for performance improvement, you help support more productive employees and a more profitable organization

What to do:

- ☐ Monitor resources to ensure they are allocated according to the project plan. If not, how could the problem be resolved?
- ☐ Consider how much time you spend observing and discussing issues with others.
- ☐ Identify an interpersonal skill that someone you work with does well and compliment the person.

Other actions:

- ☐ ___________________________________
- ☐ ___________________________________
- ☐ ___________________________________
- ☐ ___________________________________
- ☐ ___________________________________

Remember

- ✓ Monitoring activities provide insight into when resource availability may be in question, and controlling activities facilitate the corrective actions needed to deal with problems.
- ✓ Being assertive and cooperative is the best way to resolve conflict, but it is not always easy, especially when emotions run high.
- ✓ Training can increase productivity over 20 percent, but training and coaching combined can result in nearly a 90 percent increase in productivity.

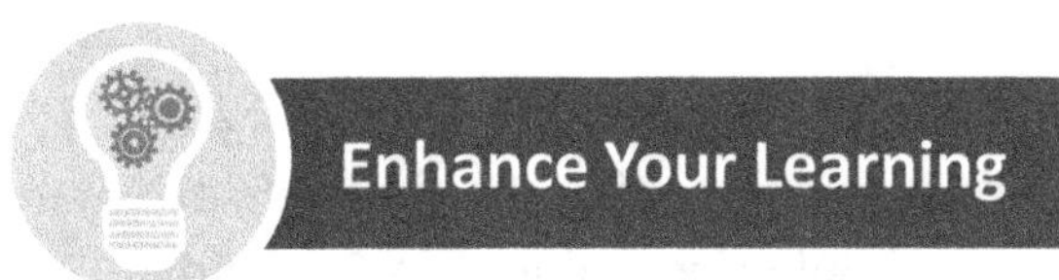

Enhance Your Learning

View the following six-minute video by ProjectManager about conflict resolution.

Whitt, J. (2013). *Conflict Resolution Training: How To Manage Team Conflict In Under 6 Minutes!*

Available at:
https://www.youtube.com/watch?v=PHJ8eybXJdw

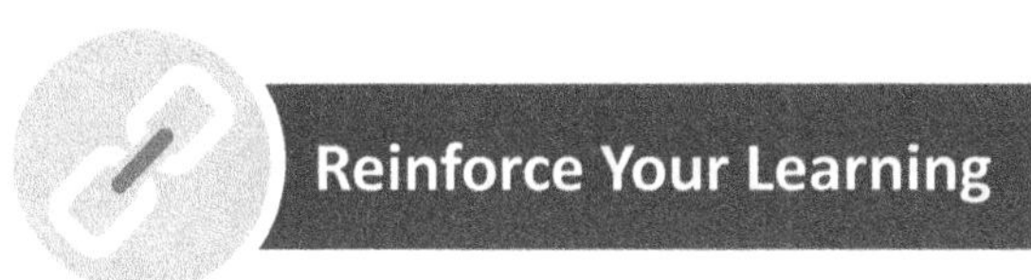

Reinforce Your Learning

The Project Team. A project team member who reports to you describes a project meeting where conflict arose because some people were unable to work the planned amount of time due to unexpected vacation scheduling. This led to an adjustment in the project's original schedule baseline. The meeting minutes noted "resource unavailability" as the cause.

What are your initial thoughts about this sequence of events?

What tools did we discuss that you might use in this situation?

What would you do?

Coaching Worksheet: Coaching Session. Assume that you plan on conducting a coaching session with the people whose unexpected vacation required adjusting the project's original schedule baseline. To prepare for the coaching session role play, complete the Coaching Worksheet below. You will also find a blank worksheet in the Appendix, Part C.

Coaching Worksheet

Coaching Situation: ______________________

Communication and Problem-Solving Steps:

1. Describe the problem in a non-confrontational way.

2. Ask for the team member's assistance in solving the problem.

3. Brainstorm causes of the problem or contributing factors.

4. Identify possible solutions.

5. Agree upon a specific action plan.

6. Set a follow-up meeting.

Identify Components of Risk Planning and Management

Risk is a significant element of any project and **can be found during all phases** of the project management life cycle. In the initial planning phase of a project, the original scope, cost, and schedule baseline plans describe the project risks and also the contingency plans to mitigate the impact on the project if any of the risks were to happen.

In the project's execution phase, the project manager must **monitor and control** the identified risks, while also being aware that new issues and risks will arise as changes to the project occur. **Unidentified risks** that arise often result in **workarounds**. Changes may impact the scope, cost, and schedule baselines, and project managers must not only evaluate how the changes affect the project, but also consider the changes from a risk perspective.

Risk Management Tools and Techniques

The **risk register**, **risk mitigation** and **contingency plan**, and the **risk breakdown structure** are three foundational references that project managers use to determine the current risk in a project. Risk assessment and risk audit are somewhat related, and we will address them together later in this concept. Project managers use these references to evaluate the original risks, as well as any new risks identified when project changes are introduced.

You can review a **risk register**, illustrated below, to determine if any expected risks have occurred. It highlights the most important risks according to ratings defined by the organization or project. In our example, "resources not available" is identified as a critical risk. Pay the most attention to critical risks. As part of monitoring and controlling risks, you can also assess the planning effort by reviewing the status during the execution phase.

Risk Register Example

Ref. #	Description	1. Probability	2. Impact	3. Probability Score	4. Impact Score	5. Severity Score	6. Severity Description
1	Resources not available.	H	H	5	5	25	Critical
2						0	
n						0	

Working with a risk mitigation and contingency plan, illustrated below, we can determine if the risk has occurred or not. Then we will need additional information, such as why the mitigation plan did not prevent the risk from occurring. Was it because the plan was ineffective, or did something about the risk change to make the mitigation plan less effective?

In this example, the project manager would start by determining whether the contracted resources were put on retainer as the mitigation plan requires. If so, then why has resource availability negatively impacted the schedule? Could it be that the retained resources did not have the necessary skills, or was the problem something else?

A secondary question would be whether or not the contingency plan worked. If the shortage of resources had to do with skills outside of the retained resources' capabilities, then there is no way to engage the contracted resources as stated in the contingency plan. The mitigation plan was flawed because it did not identify the right resources to be retained.

Risk Mitigation and Contingency Plan Example

Ref. #	Risk Description	Severity	Status	Mitigation Plan	Trigger	Contingency Plan
1	Resources not available.	Critical	Red	Have contracted resources retained.	Calendar review, resource unavailable.	Engage contracted resources.

This may also be a good lesson learned from the standpoint of breaking the types of resources into more finite groups, which in this example would have made the mitigation and contingency plans more useful. It may also have helped the project manager to better define the critical skills needed for the project.

The other references for **risk assessment** and **auditing** are the **change request log** and **issues log**. You can use these logs to discover whether any risks introduced because of an issue or change request have been added to the risk register and analyzed for how critical the risk is to the project. If the risk is determined to be critical, then appropriate mitigation and contingency plans need to be put into place.

For an audit, the same type of review, as described above, could be conducted at a later date to determine whether the risk did occur and whether the mitigation and contingency plans worked as expected.

Variance and trend analysis is a technique project managers use to help them track the number of risks that have occurred and their impact on the project. Risks can be recorded through the change request process by noting that a change request is being generated because of a risk not identified in the planning phase.

For example, the number of change requests caused by risks may be exorbitant if the number is more than a predetermined percentage of the total number of change requests. If the risk assessment had been done correctly, you would expect a small number of risk-related change requests within all change requests.

One reason for monitoring and controlling this information is to determine, as early as possible, whether the risk assessment is broken or flawed. If so, you may decide to do a full-blown reassessment of the risks associated with the activities not yet completed. You want to avoid continuous surprises from risk-related change requests throughout the rest of the project. Change requests are disruptive to the flow of the project; too many can generate uneasiness among the project team because the scope keeps changing.

Technical performance measurement is another technique for managing risk. It can take many forms. As an example, assume that one of a project's deliverables is a reduction in the average wait time of system users by 20 percent or more. In this case, the team might measure computer server processing rates, and there are a few other things we would need to do.

One is to capture user wait time before the project and use it as a benchmark. Then we must capture user wait time after the product is ready for testing. This is the only way to ensure that the basis of success is quantitatively derived using "before" and "after" data. Do not make this decision by asking qualitative questions like, "Do the users *think* the system is running faster or slower than before?"

Reserve analysis is a risk management technique that helps project managers assess whether the monetary reserves set aside for contingency purposes will suffice, given the recent history of the project. Consider an extension of the discussion above about risks causing more change requests than anticipated. One metric impacted by change requests is cost.

As more and more change requests are approved, the consequences of the changes may consume the reserves. Assume the cost reserve created was 15 percent of the total cost of the project. As each change request is added, its projected cost must be deducted from the reserve. It could have been predetermined that once the cost reserve reached a certain threshold, it would require additional levels of authorizing entities to approve.

What to do:	**Other actions:**
☐ Review the issues and change request log examples and consider how risks are tracked.	☐ ______________________________
	☐ ______________________________
☐ Review the risks in your project and consider what tools you could use to improve risk management.	☐ ______________________________
	☐ ______________________________
☐ Review your project's monetary reserves and consider recent project performance. Is the reserve likely to be adequate?	☐ ______________________________
	☐ ______________________________

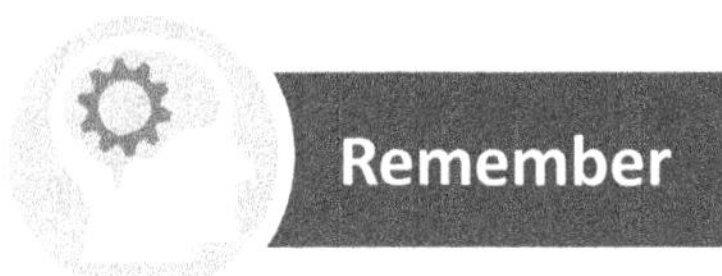

Remember

✓ The risk register, mitigation plan, and contingency plan are useful resources for risks monitoring and controlling activities.
✓ The issues and change request logs are useful resources for risk assessment and audit.
✓ It is helpful to have quantitatively supported project results whenever possible.

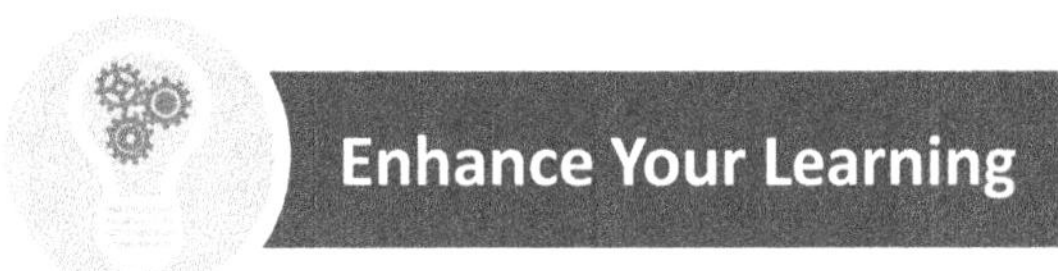

Enhance Your Learning

View the following 3-minute video by Simplilearn about identifying, monitoring and controlling risk.

Simplilearn. (2012). *Monitor and Control Risk	Project Risk Management.*		Available at: https://www.youtube.com/watch?v=TdWd_4glh_0

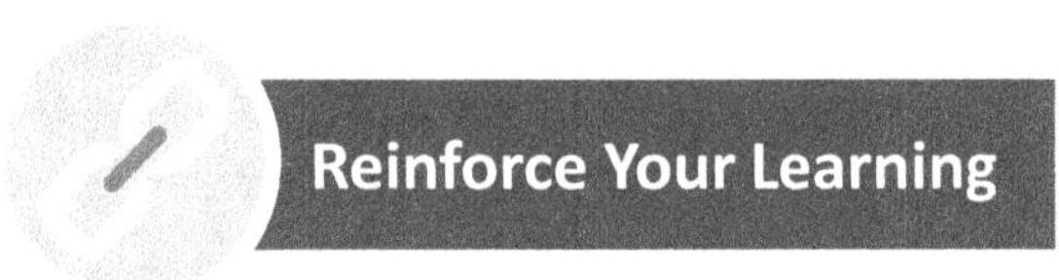

Reinforce Your Learning

Your New Home: Revisited. Recall the new home exercise from earlier where you are building a new home, and after the work had already started, you decided to add a garage to the project. In the previous exercise, you were asked to identify the part(s) of your project management plan that would be impacted by adding the garage.
Using your response(s) from that previous exercise, what **risk management tools** and **techniques** could you or the contractor use to effectively monitor and control risk in your project? Explain how you would use these resources.

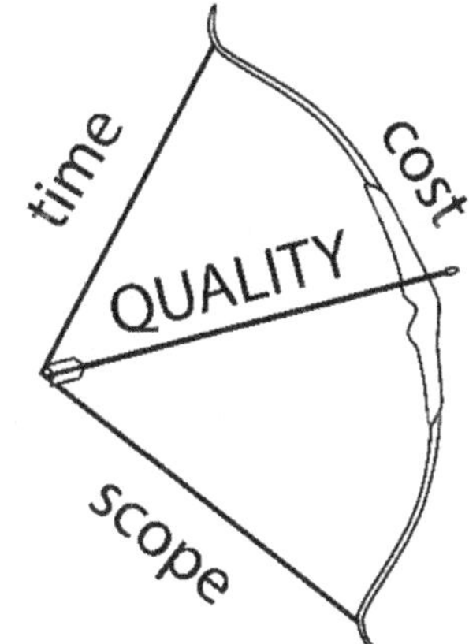

Review Quality Monitoring and Control Tools

You either get better or worse. These words are as true in business as they are in sports. In today's business environments, work systems and processes have received a great deal of attention because they are central to judging and managing quality.

Quality is a variable and can only result when **variation** is **understood** and **controlled** as part of project management. **Quality assurance** relates to how a process is performed or how a product is made, **quality control** is more the inspection aspect of quality management.

Quality **cannot be inspected** into products and services. The "find it and fix it" approach to quality is ineffective.

How you monitor and control quality is primarily determined by the kind of processes in a project environment. In environments that *do not include* repetitive processes, quality monitoring focuses on **performance measurements** or the **metrics of the project,** including scope, schedule, and cost. In environments *that include repetitive processes*, additional quality monitoring and control are needed.

In this concept, we will discuss enterprise environmental factors and organizational processes to better understand what elements of quality must be included in a quality management plan. We will also review some tools and techniques project managers typically use to monitor and control quality. These are metrics, quality checklists, work performance data, change request logs, and issues logs.

Process Management

A major management strategy for quality is problem **prevention** through the application of statistical quality control and process management before and during an operation.

Process management is nothing more than good business management, and good business management involves problem management. Problem management entails:

- Organized identification of problems
- Formal improvement of quality and processes by solving problems
- Planned avoidance of problems

Baseline Performance and Entitlement Performance

Each function in an organization performs at some measurable level that is often referred to as baseline performance. Examples are unit cost, cycle time, and time to complete new designs.

Organizations make investments in equipment, software, designs, processes, and people. These investments entitle an organization to an "as designed" level of performance, most often substantially higher than baseline. In this sense, entitlement and optimization are synonymous.

The following are early pioneers in the quality/continuous improvement journey:
- Shewart: Statistical Process Control charts (SPC)
- Deming: Total Quality Management (TQM); Plan-Do-Check-Act
- Juran: fitness of use; 85 percent of quality problems are under management control
- Crosby: zero defects; quality is "free"
- Ishikawa: cause and effect diagram (fishbone); quality circles
- Taguchi: design of experiments
- Galvin, Gumpta, Harry, Schroeder, Zinkgraf (Motorola): Six-Sigma
- Ohno (Toyota): Lean; Total Production Systems (TPS)

Quality metrics vary from project to project. We will walk through a set of metrics associated with the Making Copies example you are familiar with. These metrics are related to the improvement of time to produce copies and the reduction of their cost. This is relevant in most environments because valuable resources can be wasted on something simple such as making copies or recovering documents sent to the printer.

Quality checklists are customized to a situation. The intent is to remove as much variation from an activity or process as possible.

For example, below is a sample checklist for making copies. The purpose is to ensure that everything at the beginning of the process is correct. This should eliminate some of the things that previously introduced variation to the process.

Making Copies Checklist Example

Copier Checklist:

___ Copier is clear and available.

___Previous job documents not present

___ Control panel is correct.

___Set to appropriate number of copies.

___Set to the correct paper size.

___Set to correct simplex/duplex setting.

___Green light appears for making copies.

Work performance data considers how frequently problems occur. In our making copies example, we collected data using a check sheet, as illustrated on the page that follows. It shows the most frequent copier problems, and how much work time is lost because of those problems. Examining this data may also reveal additional problems not initially identified in the planning process.

A **change request log**, as we discussed earlier in the course, is maintained to ensure that any ideas for improvement (change requests) are logged for follow-through and control. Part of the follow-through is in the form of the request actually being fulfilled. The other part is validating that what was done to satisfy the request actually satisfied the request.

Deliverables include anything to be delivered within the scope, or the change requests, of the project. Not only should a deliverable be put in place, but it should be done correctly. After the deliverable is in place, you should verify that the anticipated benefits are being realized. For example, as we progress toward optimization of the copier process, all improvements that are discovered and implemented will be verified to ensure the gains are realized and the anticipated benefits validated.

Tools and Techniques for Managing Quality

Tools and **techniques** for monitoring and controlling **quality** include:

- Five basic quality tools
- Statistical sampling
- Inspection
- Approved change request review

The following is a discussion of **five commonly used quality tools**: check sheet, pareto diagram, cause-and-effect diagram, affinity diagram, and process flow map.

Check sheets help you determine what types of problems, or defects, are occurring and how frequently. A check sheet that addresses the defect or problem of spending too many resources making hard-copy documents is illustrated below.

In conversation with stakeholders, the check sheet was put at the copier so users could record appropriate information while going about their daily work. This version of a check sheet includes the amount of time lost for each occurrence to determine how much of the organization's resources is affected by this process. The actual format of the check sheet is determined by the situation.

The check sheet example below is the result of a project deciding to record issues, or defects, types, minutes lost, and the number of occurrences.

Making Copies Check Sheet Example for Recording Issues (Defects)

Defect Types	Time Lost (minutes)	Mon	Tue	Wed	Thu	Fri	Total
Copier Checksheet							
Printer is down.	100	1	2		2		5
Printer jam, fixed.	20	2		2		1	5
Printer jam, someone else fixed.	30		1	1	1		3
No paper, had to load.	85	2	4	4	5	2	17
No paper, none available	50			1		1	2
Ink, had to load.	25	1					1
Ink, none available.	60			1			1
Total	370	6	7	9	8	4	34

Please record time lost for each occurrence.

Using this information, we can produce different versions of a **Pareto Chart**. The first version we will look at is the number of occurrences with each defect type. This enables the team to focus on defects that occur most frequently. Addressing the most frequent occurrences will typically give you the biggest bang for your buck.

The illustration below is a Pareto Chart for the most frequently occurring defect.

As you can see, Load Paper is the most reoccurring defect by far. It is understandable why this activity would occur regularly.

There could be some time spent to find a more efficient method of loading the paper, especially if a new method would reduce the chances of a paper jam.

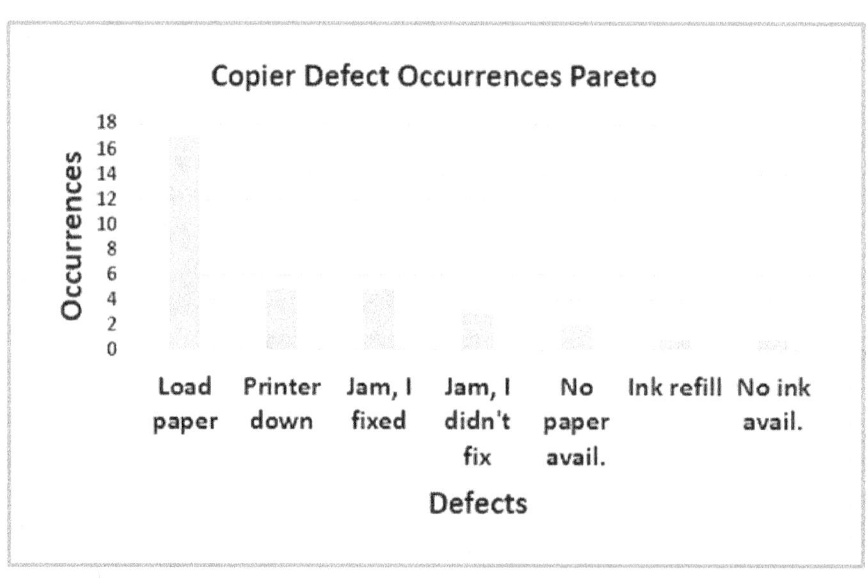

If we were to look at the data differently because **the problem is the amount of time and money put into producing hard copy**, we can take the same data and **create a pareto chart for the time lost in dealing with the defect.** We now have a new number one defect, which is Printer down.

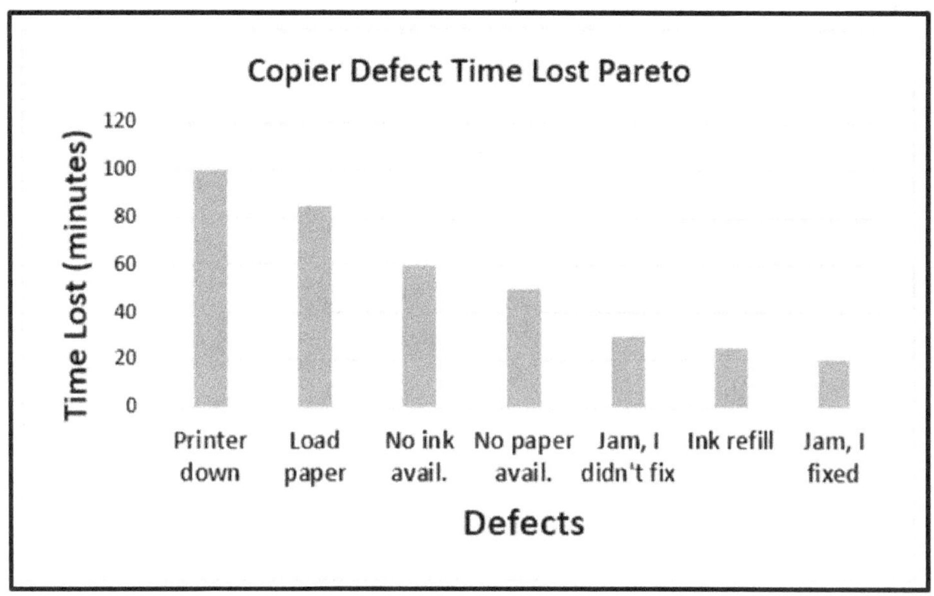

In our copier scenario, the project team had already created a **cause-and-effect diagram** (also known as an **Ishikawa** or **Fishbone** diagram), shown below, during the planning phase of the project. We can use that diagram to validate that all root causes have been identified and decide how we might best take advantage of that prior work.

Copies Cause-and-Effect Diagram Example

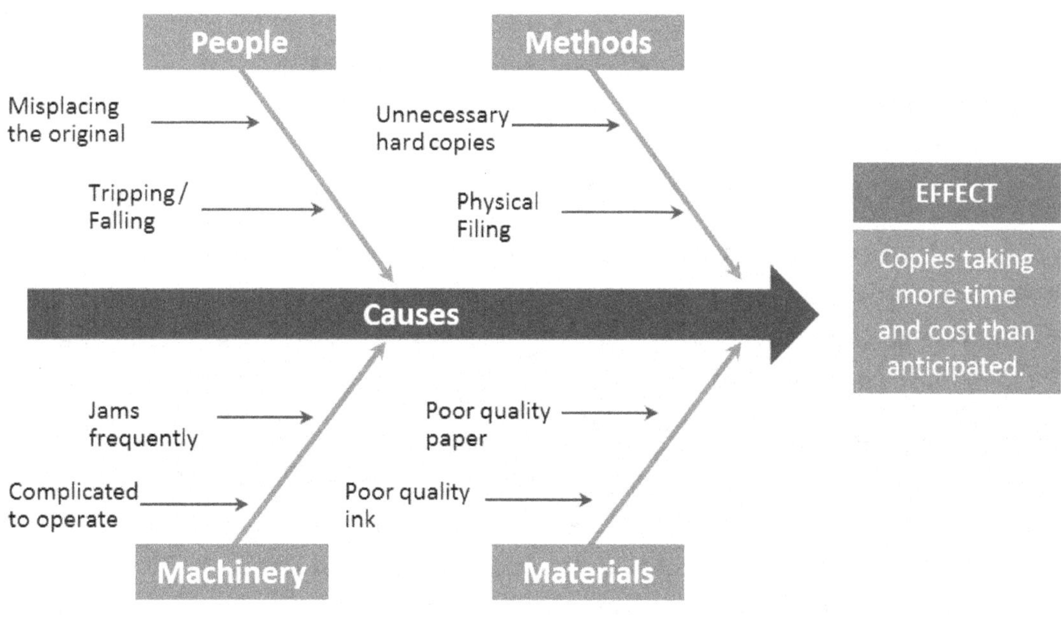

Understanding what connects activities or the assumptions behind some activities can be helpful. These are methods (procedures people follow), as opposed to machinery concerns.

In our example, we will consider copying activities that result in unnecessary hard copies and physical filing root causes. We will use an **affinity diagram** to think through some considerations associated with these activities. The affinity diagram is a method which can help you gather information and data and organize them into groups or themes based on their relationships.

We will start by identifying the related activities of making a hard copy, filing a hard copy, and distributing a hard copy, as illustrated in the affinity diagram that follows.

Affinity Diagram Example

Making hard copy:

- Using the copier is painful. There are always issues.

- I spend at least 5 hours a week making hard copies.

Filing hard copy:

- I never access filed hard copies for my reference.

- I never access filed hard copies for others' reference.

Distributing hard copy:

- We assume other people actually use them.
- No feedback about timing.
- No feedback about content.
- No feedback about format.
- No feedback when I was out of the office.

The **making hard copies** discussion brought out a couple of thoughts. First, using a copier is typically not a user-friendly experience. It often leads to frustration because of the problems encountered in trying to use it. We have initially addressed this issue from the standpoint of analyzing the problems with making copies and what would be needed to fix those problems or minimize their occurrence.

The **distributing hard copy** discussion brought up a few ideas. The person doing the distributing was just assuming that recipients were using them—but never got feedback about the timing, content, or the format of hard copies. You would expect compliments or, at the very least, complaints, unless, maybe, no one is actually looking at them . . . .

The discussion related to **filing hard copy** led to the idea that there was no apparent need for future reference. The person making and filing the hard copy never referenced it later, not even by request from anyone else. There are environments where hard copy must be retained for some time. This project was not one of those environments. The filing was done only because it was a common practice in the organization.

This line of discussion may seem almost too simple from a quality perspective. It is common for one problem *(resources wasted during the copy process)* to challenge something upstream, or downstream, that may dramatically change the direction of the path forward. It is also common for an outside pair of eyes with no history of how things have always been done to ask questions that may stir insights from those with a working knowledge of the process. This can open doors for changes that improve the process.

The checklist discussed earlier addressed the machinery portion of the root causes. The methods merit more discussion because they are a procedural issue. If the team is making unnecessary hard copies that represent a significant amount of volume for the copier, we want to address this issue first because resolving it could also resolve the issue of hard copies taking unnecessary resources. This could also spill over into the physical filing issue because if hard copies are of no value, then there is no value in filing them.

Process flow maps are an interesting way to represent current activities and also what the flow could be in the future.

Below is the original process *(current state)* for the making copies process. It considers the person doing the job and those who receive hard copies.

Original Process for Making Copies

Gather — Scan (if needed) ⟶ Save to network drive ⟶ Email distribution (if needed) ▸ File hard copy

Below is the agreed-on improved process (future state) for making copies. The discussion included guidelines on what types of documents should be scanned and the email distribution.

Improved Process for Making Copies

Scan (if needed) ⟶ Save to network drive ⟶ Email distribution (if needed)

We have reviewed some tools used for quality monitoring and control. Our example describes how working through a process using these tools can change the way things are done. The results of using one tool may completely negate the results of another tool. In our example, the Copier Defect Check Sheet and resulting pareto charts were negated by the determination that the volume of hard copy being produced was unnecessary. That decision alone will significantly diminish the volume of hard copy and the resources needed to handle them. Since resource consumption was the "effect," the results of our exercise are successful.

Statistical sampling and inspection are tools that can be helpful for processes that create a large number of outputs, such as an assembly line. Several methods of sampling the outputs every so often and inspecting them for defects are discussed in common Six-Sigma and other statistics textbooks.

<table>
<tr><td>

What to do:

- ☐ Review a cause-and-effect diagram for your project and validate it.
- ☐ Review the quality management plan for specifics of monitoring and controlling quality.
- ☐ Review a process flow for your project for potential improvements.

</td><td>

Other actions:

- ☐ _______________________
- ☐ _______________________
- ☐ _______________________
- ☐ _______________________

</td></tr>
</table>

- ✓ Quality considers aspects of the project and the organization's processes.
- ✓ Planning documents are instrumental for use in monitoring and controlling quality.
- ✓ Quality monitoring and controlling tools need to be customized to the situation.

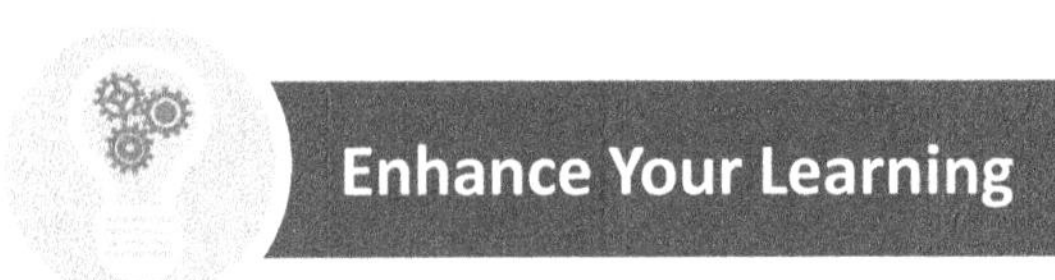

Watch the following 3-minute video by the Arizona Western Region Public Health Training Center to learn more about affinity diagrams.

<table>
<tr><td>

Andrade, R. (2013). *Affinity Diagrams.*

</td><td>

</td><td>

Available at:
https://www.youtube.com/watch?v=jvTSsJrDZec

</td></tr>
</table>

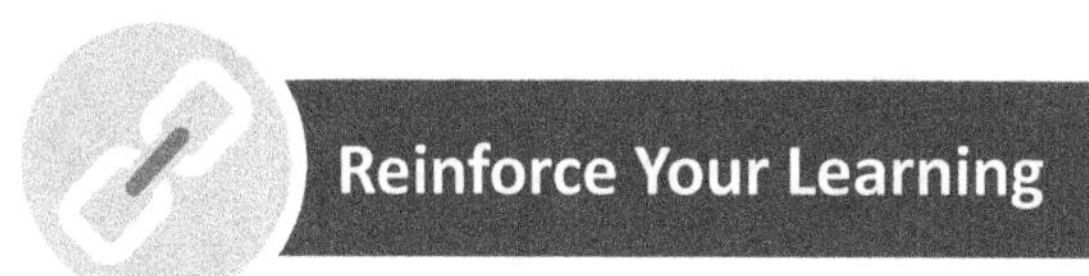

Affinity Diagram. Pick a complex issue such as the "**Construction of a House**" or another project of your choice. Start by identifying two to four processes or ideas, as indicated in the affinity diagram below. Then ask the team to brainstorm ideas associated with those processes or ideas. The rationale of the ideas is not an issue. It is enough to just gather all the ideas. Once everyone has time to contribute, organize the ideas into groups. The groups usually become apparent, but you may need to make a few judgment calls.

This exercise can be done using the template below. Or you can use Post-it-notes to capture the team's ideas and then move them into the appropriate affinity groups on the diagram that follows.

Affinity Diagram Exercise

Process or Idea.

- Supporting ideas.

-

-

Process or Idea.

- Supporting ideas.

-

-

Process or Idea.

- Supporting ideas.

-

-

Once all the ideas are on the diagram, review the quality management plan and the project management plan to determine whether everything in the affinity diagram has been accounted for/addressed. If not, recommend appropriate adjustments.

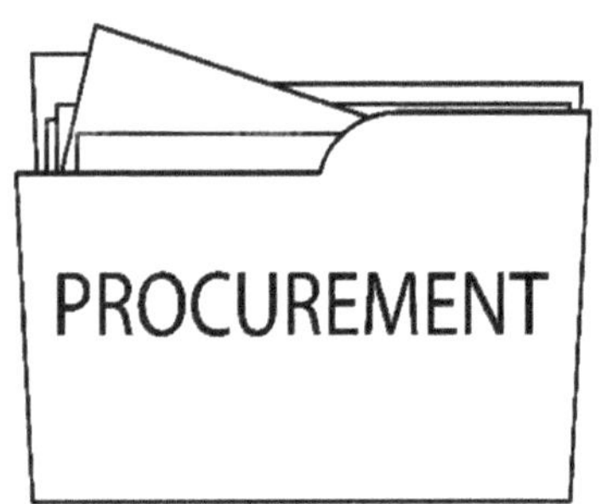

Monitor and Control the Procurement Process

Procurement refers to the process of obtaining goods and services, and so it is part of project management. Project managers usually work with others in an organization, such as the procurement, contracting, or legal departments, to obtain the goods and services needed for a project. At a minimum, project managers must know how to negotiate and manage contracts.

Project managers are also responsible for putting in place monitoring and controlling mechanisms to ensure reliable and effective procurement practices during the life of the project, and to see that goods and services are obtained as specified in the project plan.

In this concept, we discuss ways to keep procurement performance in line with projections.

Tools and Techniques for Procurement Monitoring and Control

The following are some of the commonly used tools and techniques for procurement monitoring and control:

- Contract change control system
- Procurement performance reviews
- Inspections and audits
- Performance reporting
- Payment systems
- Claims administration
- Record management system

A contract and change control system is often needed on large complex projects and is typically included in the change control management plan. Depending on the size of the project, this could be an additional section, or particular mention, for each contract, and it could be general guidelines for contract changes.

Each contract can have specific guidelines that justify a separate paragraph or special mention. For example, if the procurement has to do with the delivery of materials to be used in the production of goods, the guidelines may be different from a procurement contract for technical services.

The specific references or general guidelines can also define what thresholds require what types of reviews and signatures.

Procurement performance reviews should be included in the quality management plan. This would include performance requirements with specific quantification of acceptable and unacceptable performance levels. In the case of materials being procured to produce a finished product, you would need to establish procurement standards that the supplier will be measured against.

As an example, assume a project to build a new outdoor obstacle course. The materials needed are wood for the scaling walls, and other large objects to crawl over and under. The grade and texture of the wood were specified, and the project schedule agreed on.

One month into the project, the supplier advised the project manager of production problems because of a problem getting the raw materials. This looked as though it would cause the schedule to slip.

Stipulations could have been built into the change control/performance parameters that required the supplier to either find another raw material source or another supplier to provide the materials within the published project schedule. However, if there were no such provisions, then the project manager may have no option other than waiting for the material and adjusting the schedule accordingly.

Inspections and audits are also part of the quality management plan and procurement management plan. To extend the previous obstacle course example, it turns out the original supplier did return to production and shipped the materials late. Upon receipt of the materials, the project team discovered they were not finished to the specifications and rejected them because course participants may have been injured from the rough surface of the materials. Thus, the schedule suffered an additional setback because of this issue.

Depending on the agreement associated with inspections and audits, it may be determined that the product is replaced or that the supplier comes to the site and corrects the defect there. These options would be agreed upon as part of the inspection and audit supporting details.

Performance reporting can be managed in several ways. One approach is a procurement scorecard, as illustrated below. The scorecard would include the essential data associated with supplier performance. The type of performance shown in the example would be cause for alarm and corrective action.

Corrective action could be in the form of reimbursement of some cost, going to a different supplier with some funding from the original supplier, or some type of monetary penalty. All potential options should have been discussed before the procurement contract was signed.

Procurement Scorecard (reporting) Example

Procurement Scorecard – Supplier Y					
Date	Event	Cost	Quality	Timing	Comment
2/3/yyyy	Missed delivery date.	1	N/A	2	Delay caused by raw material supplier.
3/6/yyyy	Materials not to specification	1	2	2	Inferior raw materials caused inferior finished materials.

n...

Legend: 1=Acceptable, 2=Unacceptable, N/A=Not applicable

The idea behind implementing a procurement system is that once implemented, it can streamline the procurement monitoring and control efforts. Such a system would include a **payment system**, **claims administration**, and a **record management** system.

A procurement system should provide a ready reference to overall procurement status at any point in time and a color-coding system is often helpful.

If business rules are established that require a specific corrective action for a rejection, these can be built into the system to ensure that appropriate steps are followed. This information would also contribute to the development of the procurement scorecard example above, but because all information is now in one system, the scorecard is an electronic by-product of the information being accurate and current.

This same information can then be rolled up to the procurement contract number level to show the current status of each. This enables monitoring and controlling at the procurement contract number level, as illustrated below.

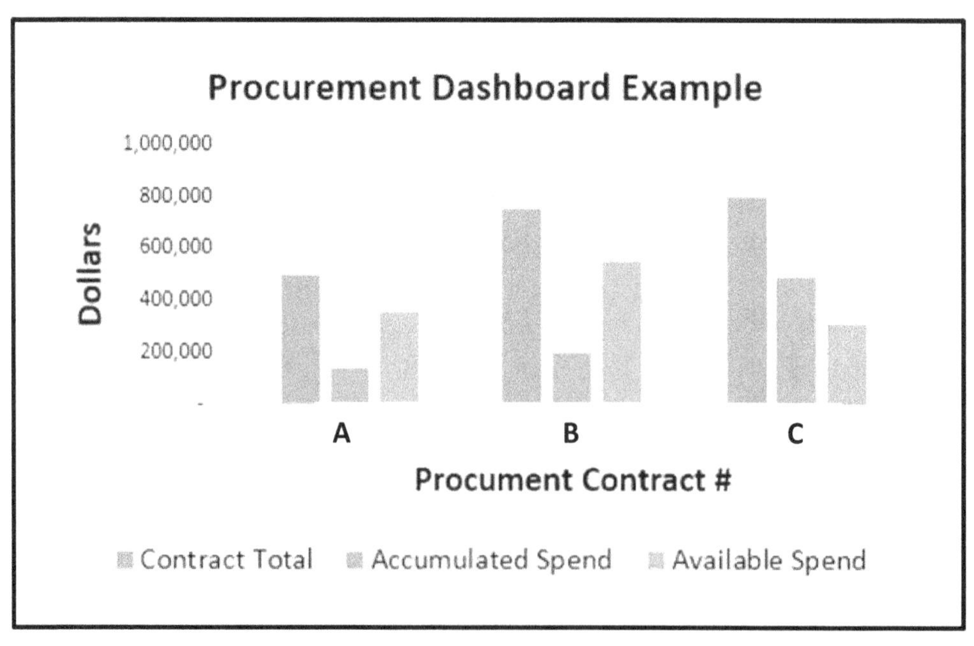

Project managers use the **dashboard to gain insights** related to procurement performance and to help determine if current procurement costs make sense compared to projections.

For example, if a particular contract was expected to have the most spending in the latter part of the project, and the project is only in its early phases now, but the contract spend is higher than anticipated, the project manager would want to find out why. If it is because planned purchases were made earlier than expected because some predecessor tasks were completed sooner than anticipated, that would be acceptable.

If, on the other hand, the amount of the earlier spend was higher than what was anticipated, there should be a review with the supplier to determine how this will be handled. If the contracted rate was based on the current market price with the anticipation that it would not increase at the rate that it has, then the project manager may need to use reserves that should have been set aside for the risk associated with the price being variable.

If the price was fixed in the contract, but the supplier is now charging more because of variables not included in the contract, that also needs to be addressed. In other words, let the supplier know you plan on paying only the contracted price.

You can also add business rules for when a contract spend is reaching a certain percentage of the total. For example, the project manager may want to know if the current spend total has reached a higher percentage than the calendar date progression of the project. This could apply to a service contract where you anticipated the same amount of service hours equally for each month. At the end of the first month, the billing might be, for example, 20 percent of the total contract when you were anticipating 16.6 percent.

If the overspend was because of additional work completed, and if the next month's billing should be less and offset the difference, then all is well. If the first month of work is taking more effort than anticipated, and it looks like this may continue, additional actions might be needed. These could include using procurement reserves, renegotiating the rate with the service provider, or possibly re-baselining the procurement plan.

What to do:

☐ Review the contract change control system that applies to your project.
☐ Verify the procurement activity that is associated with your responsibilities within the project. Are they on track with the original baseline?
☐ Check to see how suppliers are doing on your project's procurement scorecard.

Other actions:

☐ ___________________________

☐ ___________________________

☐ ___________________________

☐ ___________________________

☐ ___________________________

Remember

✓ Procurement systems can enable monitoring and control of the project's procurement process.

✓ Procurement scorecards are an excellent way to rate supplier performance.

✓ Variations from the procurement baseline should be dealt with promptly.

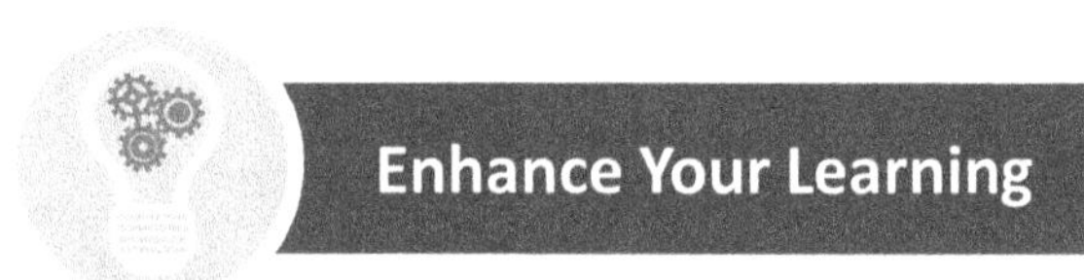

Enhance Your Learning

Watch the following 4-minute and 5-minute videos from ProjectManager to learn tips for project tracking and monitoring. Consider how you can apply this information in your work:

Deen, D. (2011). *Top 5 Project Tracking & Monitoring Tips for Project Managers.*		Available at: https://www.youtube.com/watch?v=CTg_9K5K6hE

Bridges, J. (2019). *How to Monitor Daily Progress as a Project Manager.*		Available at: https://www.youtube.com/watch?v=E2Qahnv1j1g

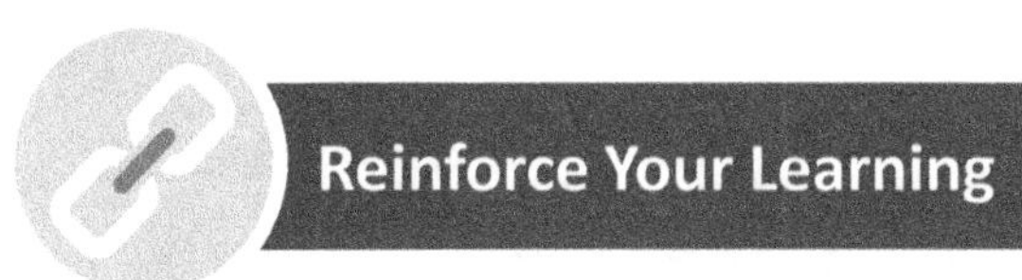

Procurement Dashboard. The following is procurement information for a six-month project.

- Contract # A23454 is for equipment purchases and has a majority of the spending in the first part of the project.

- Contract #A24567 has a majority of the spending halfway through the project because of supporting activities for new equipment purchased with contract #A23454.

- Contract #B23435 is set to spend equally across all six months of the project.

Below is the procurement dashboard report created at the end of the project's third month. Assume you are the project manager and consider the questions that follow.

Procurement Dashboard Exercise – 3 Months			
Contract #	Contract Total	Accumulated Spend	Available Spend
A23454	900,000	154,000.23	745,999.77
A24567	100,000	10,000.00	90,000.00
B23435	600,000	425,345.23	174,654.77

1. What observations can you make based on the information provided?

2. What types of things or events may have contributed to what you observed?

3. Can you identify a procurement management tool or system that could be used to help with early detection of the problems?

276

I am prepared for the worst but hope for the best.

—Benjamin Disraeli

Monitor and Control Project Communication

Communication has a significant influence on whether projects succeed or fail. It is no exaggeration to say that effective and timely communication throughout the life of a project is crucial to its ultimate success.

The execution phase of the project is often a dynamic period in the life of a project. For example, you may discover differences associated with the original activity projections. These could be activities that were overlooked or, for example, risks with underrated impact, not to mention change requests that affect the scope.

All such situations require an analysis of the effect on the project's scope, cost, and schedule. Several people could be involved in the analysis, and after decisions are made, you must communicate the results.

During the planning of the project, the project manager must define the **reporting structure** and communication chain, both for people internal to the organization and any who may be external.

The outputs for communication monitoring and control typically include things such as:

- Project communications in all forms
- Work performance information
- Change requests
- Updated project documents

Project Manager Responsibilities

The project manager is typically the focal point for communication with the client/stakeholders. To avoid communication problems, the project manager should also be the only person who communicates with consultants, agencies, or governing bodies. However, there can be exceptions to the rule. If your project has exceptions, document them at the beginning of the project.

Besides establishing clear lines of communication and ensuring that everyone is informed about what they need to know, the project manager must also consider time management for stakeholders and the project team. In this day of electronic communication, it is easy to send messages, memos, and reports to many people at one time. But too much communication going to people who do not need the information can be a significant source of wasted time.

Sometimes the opposite happens, and a project manager might not distribute communications to people who need them. For example, based on a change in a project, the project manager may forget to add a new team member to a communication list. The result would be that someone does not know what is going on, what has been added or modified, or what may need their attention. This will lead to a frustrated team and project delays.

As a project manager, you are responsible for ensuring that the communication plan is current and that everyone has the information they need. A good habit is to routinely check with stakeholders and team members to verify they have what they need to do their jobs.

In addition to electronic and paper communication, periodic **project review meetings** are intended to communicate information, both from the project manager to the team and from the team to the project manager. Information should be communicated within the context of the project scope, schedule, and budget. The stakeholders may also attend these meetings, or you might set up separate meetings with them, depending on what works best for your project.

Controlling Project Communications

Controlling a project's communication from start to finish requires planning and discipline. Such preparation includes identifying who needs what information and when. This includes the types of information to be communicated, how it will be delivered, and the frequency.

Obviously, the complexities of communication are affected by the size of the project. For larger projects, initial questions to ask might include: How will the ongoing status of the project be communicated? Who will receive project status reports? Who needs to know about project concerns and problems? Is a project team directory available? Is a communication matrix available?

Following through with planned communication occurs throughout the project life cycle. If a stakeholder communication matrix is not available, it would be beneficial to develop one.

The following is an example communication matrix. Stakeholders are listed in the lefthand column, the types of project information needed display across the top, and the cells indicate the communication medium and frequency.

Stakeholder Communication Matrix Example

Stakeholders / Project Information	Contracts	Project Plan	Informal Updates	Design Submittal	Kick-off Monitoring	Progress Reports	Scope Changes	Meeting Minutes	Project Specific Info.	Invoices	Etc.
Directors											
Project Managers											
Client Project Managers											
Administration											
Discipline Managers											
QA Managers											
Regulatory Agency											
Etc.											

Communication Medium

1. Written Reports
2. Telephone
3. Web Page
4. Meeting
5. Email
6. Project Related Information
7. Other

Frequency:

D: Daily
W: Weekly
M: Monthly
S: Schedule
O: Occurrence

Baseline Changes

Changes to the project scope, schedule, and budget must be communicated with a comparison to the original baseline and the current baseline. This includes the most current information related to change requests, resources, costs, and risks.

We have discussed monitoring and controlling tools and techniques for schedule, scope, and cost. All of them take into account the most current baseline information. But this is based on the concept that established guidelines state when an additional baseline may be required. Include the guidelines in the project, scope, schedule, cost, and change management plans.

Timely and effective communication associated with a baseline change is a big deal. A baseline change should be a flag for the client/stakeholders and team that at least one metric has changed enough to require a new baseline.

Communicating the Differences Between Baselines

You can use a project information management system to describe the differences between baselines. Below is an example of a project's original baseline information. You can see the anticipated length, or duration, of each activity. This information is saved as baseline 1.

The red box with a 1 inside highlights an activity that will be changed because of a recent discovery. The duration, or how long it should take to complete, will be extended. The red box with a 2 highlights a successor activity to the activity with the 1 in the red box.

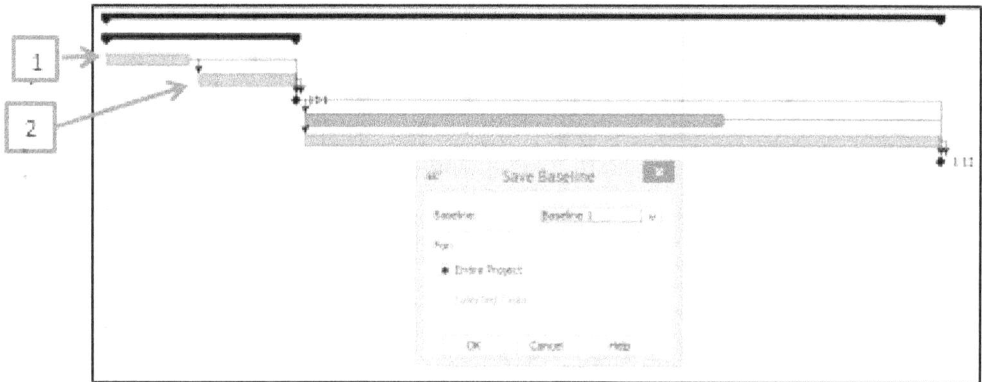

Baseline 1 Example

We add additional days to the duration column for activity.

Notice, in baseline example 2, below, the software has automatically elongated the bar for activity 1 and also pushed activity 2 to a later date (the other activity is a successor).

Also, notice that the 3 in the box highlights that the software shows the duration was in baseline 1. The software clearly illustrates the difference between schedule baseline 1 and schedule baseline 2.

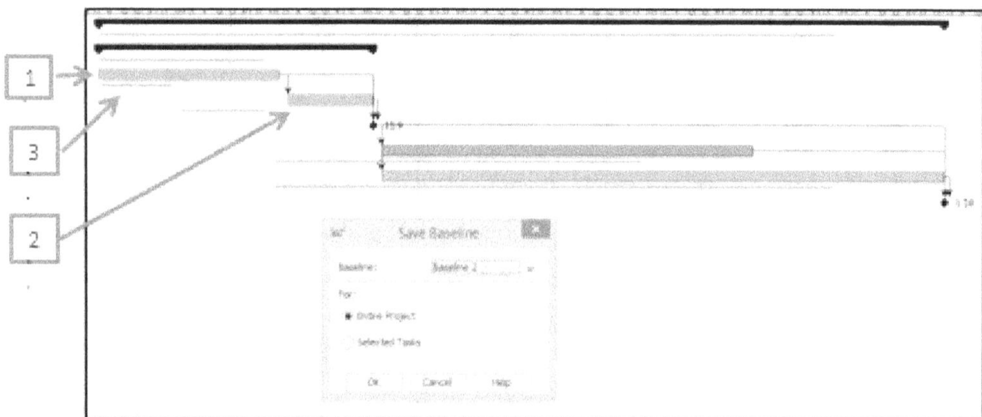

Baseline 2 Example

All purple lines show the original schedule for each activity. This provides a powerful visual that illustrates just how much the schedule was impacted. The people receiving the schedule update can see the impact and ask as many questions as needed for them to feel comfortable, they understand the current schedule.

For example, a question might be: Why was the activity elongated? As the project manager, your response might be something recently discovered was not anticipated in the original projections.

The next question might be: If that happened with this activity, how many other activities may have a similar surprise associated with them? Your response might be that the activity was more complicated than any others and that you reviewed the problem with the project team, and they are confident all the other estimates are sound.

This type of communication can be challenging, but it is necessary to keep the client/stakeholders in sync with the project, so they can make informed decisions.

When delays, or other surprises, are reported, it can degrade confidence in the project and sometimes in the project manager. An experienced project manager must do due diligence to ensure that any additional surprises are kept at a minimum. One way is to work through the root causes of each significant surprise and verify the same type of surprise is not likely to occur in another part of the project.

Clients/stakeholders consider how many surprises have been reported throughout the life of the project. They rely on their confidence in the project manager and their intuition as to whether the level of surprise is acceptable moving forward. This likely will include some discussion about this topic with the project manager, so everyone has the same expectations moving forward.

The above example discussed a schedule change, but the same concept applies to any changes introduced during the project.

Project Status Reports

One tool that you must use consistently is a **project status report**. The content and format of the report should be worked out with the project client/stakeholders in the project's planning phase, along with an agreement as to how much supporting information should be included.

Below is a sample of a typical report. We will use it as our example in the discussion that follows.

Activity List with Status Example

Name	Duration	Start	Finish	Status
⊟ Project 1	45 days	11/19/…	1/20/…	
⊟ Task 1	15 days	11/19/…	12/9/…	
Task 1A	10 days	11/19/1…	12/2/…	Red
Task 1B	5 days	12/3/15…	12/9/…	Green
Task 1 completed	0 days	12/9/15…	12/9/…	Green
Task 2	20 days	12/10/1…	1/6/1…	Green
Task 3	30 days	12/10/1…	1/20/…	Green
Project Complete	0 days	1/20/16…	1/20/…	Green

Schedule, scope, and cost are the three big hitters most likely to impact the state of a project. The status report should include them, as well as a snapshot of the risks. A risk assessment should be conducted periodically, but if a serious risk arises, add it to the status report even if its status is, "We must spend more time understanding the risk before we can report on it."

Your goal in reporting status is to report what the client/stakeholders need to know as simply and informatively as possible. One way of doing that is to use red, yellow, and green status indicators, as shown above. Using the same activity list from earlier in this concept where one activity was elongated, we have marked the status of the activity as red because we missed its original end date.

The other activities that depended on this activity to be finished were pushed out to later dates. The team assured us there will not be any other surprises with the remaining items, so we marked their status as green. This visual will also generate questions associated with the red item, so the client/stakeholders can determine if they understand the reason for the change.

What to do:

- ☐ Review your most current project status report. Is what you need to know included?
- ☐ Identify the types of communication you are receiving and determine their effectiveness. Can you suggest improvements?
- ☐ Consider the client/stakeholders on your project. Do you think they know the project's current status?

Other actions:

- ☐ _______________________
- ☐ _______________________
- ☐ _______________________
- ☐ _______________________
- ☐ _______________________

✓ Things change. Effective communication of those changes is critical.
✓ If you think you are not getting the information you need, let your project manager know.
✓ The number of project baselines can indicate the significance of changes to a project.

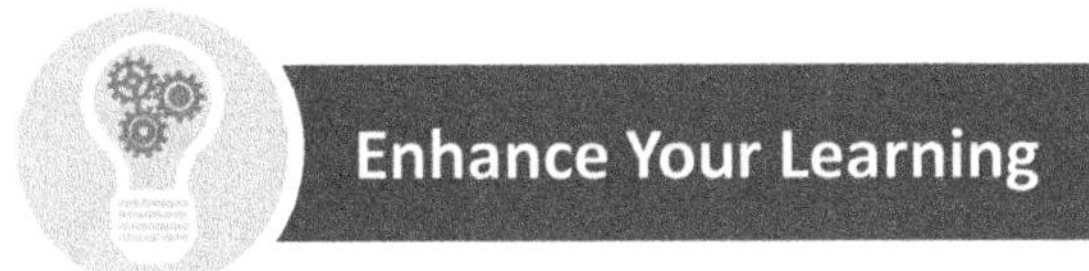

Watch the following 15-minute video to learn more about communicating with stakeholders.

Akinwale, P. (1016). *Manage Communications/Stakeholder vs ControlCommunications/Stakeholder.*		Available at: https://www.youtube.com/watch?v=ZKYmRKadujg

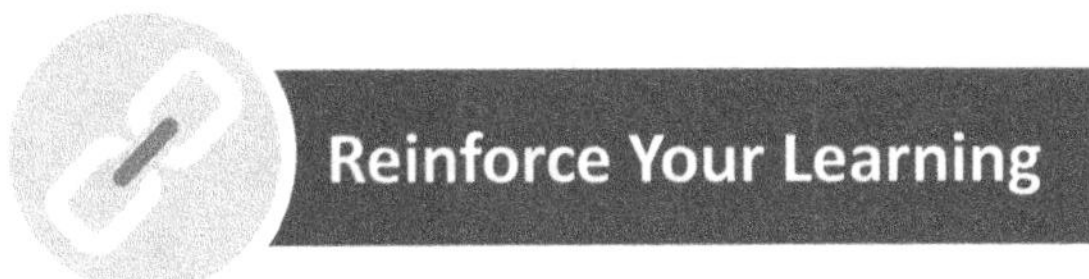

Communication Plan. Review the situation below and note your thoughts.

Jane, one of the project stakeholders, just learned about several changes to the scope of a project. The changes have caused the project to lag behind the original schedule and have increased project costs.

What forms of communications, or tools and techniques, might a project manager have used to mitigate or try to prevent this situation from occurring?

What should the project manager do to ensure that everyone is informed about project changes?

Just because you make a good plan, doesn't mean that's what's gonna happen.

—Taylor Swift

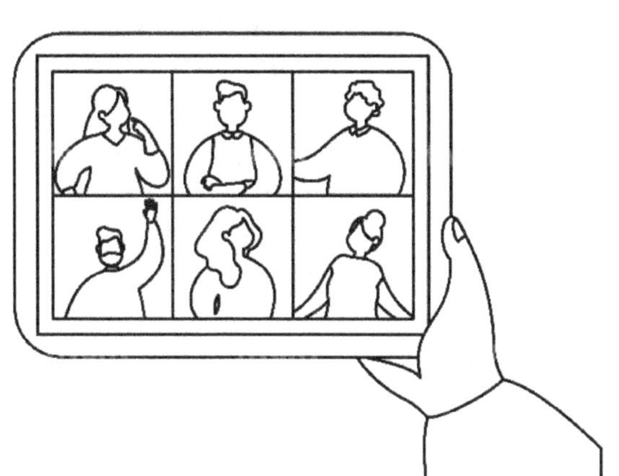

4.10

Implement Change Management Monitoring and Controlling

We will consider change management from two perspectives. The first is change management within a project; the second is change management within an organization. A project will not only experience changes as it progresses, but it will inevitably cause change within the organization. In some cases, the change can be dramatic and unsettling.

The Planning Elements Involved with Change Management

The planning elements associated with monitoring and controlling **change management** include the following:

- Project management plan
- Change management plan
- Work performance reports
- Change request log
- Issues log
- Enterprise environmental and process factors

Tools and Techniques for Managing Change

Tools and techniques for change management monitoring and control typically include:

- Expert judgment
- Change control policies, processes, and procedures, and documentation

Expert judgment draws on the experience of those most familiar with what changes. Experts help identify all factors to consider in assessing the impact of a change and how to monitor and control the change. We will walk through an example of using expert judgment later in this concept.

As we have discussed, change management within the project is primarily concerned with how things are handled when variations from the baseline occur. We mentioned using a change management plan, change control policies and procedures, change request forms, and change request logs. Another tool often integrated into larger projects is a control board.

A **change control board** helps protect a project from changes that are not advantageous. Project managers use it to help ensure that changes are incorporated into the project correctly, to ensure effective monitoring and controlling, and to oversee accountability for the results of the changes.

Change Request Example

Below is an example of how you might log a change request. While working through requirements, the team noticed there was a weakness in processing customer requests, and Jane wrote a change request aimed at correcting that problem. This improvement was not part of the original scope and will involve additional work.

Jane is the manager of this area, and she provided supporting information about why this change should be included in the scope. The change control board approved the scope change, and the project management plan has been updated accordingly.

Abbreviated Change Request Log Example

Project Change Request Log					
Request Date	Change Request Reference Number	Requestor Name	Requestor Contract Info.	Request Description	Why / Benefit
2/1/yyyy 0010		Jane Smith	Cell: 111-111-1111 Email: jsmith@org.com	Establish future state process flow for customer request processing.	Response time is too long.

The example illustrates how you might manage change. It is a format of a change request log that includes less information than the change request log we reviewed earlier in the course, and that is a template in Appendix A. You could use this format, or use the more extended version of a change request log from Appendix A.

This abbreviated example has a brief description of the change and why the change is needed, which could also be described by naming the benefit.

This is also an example of how project changes can change an organization, in this case, it will affect people's lives by changing the way they do things. This change to the project will impact three customer service representatives.

Change within an Organization

That brings us to the perspective of change management within an organization. Understanding how people do a job now and how that might change often includes mapping out the process of how things are done now (current state) and how things could be done in the future (future state).

The future state could disrupt the way things are currently done in an organization. This can be unsettling to people, particularly those who are most affected by the change. **Emotional reactions** to change are common, and **communication** is vital. As you read the list below, consider how employees might feel as they respond to the following initiatives, and the role of appropriate communication.

- New technologies
- Mergers and/or acquisitions
- Downsizing and layoffs
- Transfers or job relocations
- Leadership changes
- Reorganizations
- New policies or work rules
- New competition
- Unexpected financial or legal liabilities

The following diagram depicts some of the emotional reactions that may be experienced.

Change Process

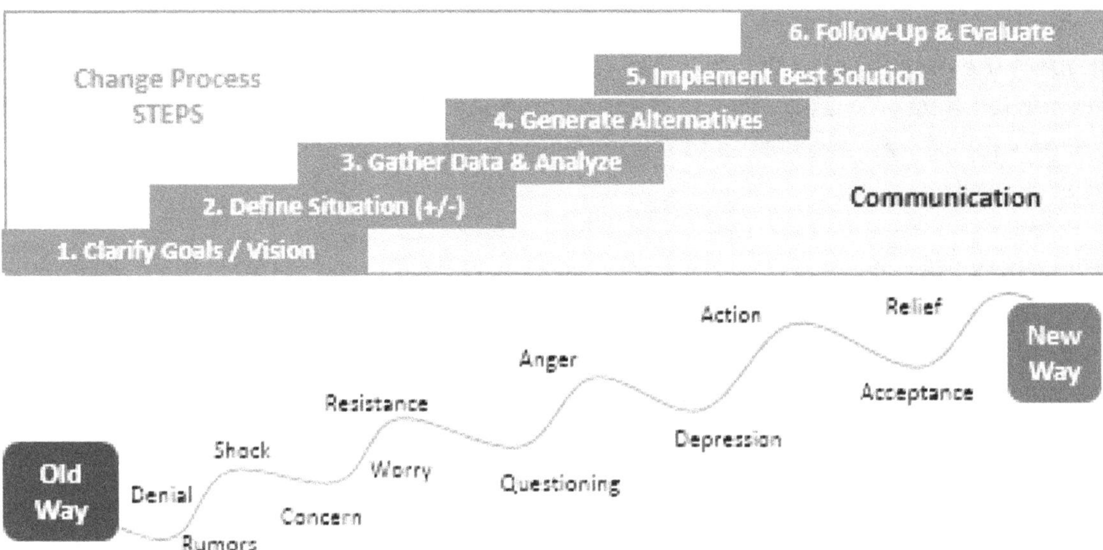

Communication about an upcoming change should be continuous. Discussions should be open, focusing on potential issues and concerns, as well as what support leaders intend to provide to facilitate change implementation. When the door is kept open for associate feedback and discussion, ambiguity is reduced while acceptance increases.

As an example, have you ever had to completely change the way you are doing something, something you may have been doing for a long time? For example, every time there is a new operating system for your computer or phone, things are not what they used to be. Something you could do easily yesterday may seem difficult today. This can be frustrating until you become familiar with the new way.

Some people may resist change, and that resistance, as well as its potential stress, must be accounted for in the change management plan. Monitoring and controlling this aspect of change management draws on soft skills.

So, back to our change request and its potential impact on the three service representatives. We need to walk through the current processes to understand what they do and how they do it. This means that you will ask a lot of questions like, "Why do you do it that way?" and "Have you considered doing it this way?" Nearly all the time, there are good reasons why they do what they do—or there used to be such reasons.

Sometimes something negative happens with a process. Corrective action is put into place so that negative thing does not happen again. Later, maybe much later, something changes upstream or downstream from that process that makes the corrective action of no value. Now, we have an activity that takes time and resources but adds no value. But this is the way that it has always been done, so we keep doing it—perhaps because we are not aware of the other change.

It is essential for those gathering the requirements, or possibly doing process flow mapping, to respect the people doing the job. They may have been doing the job the same way for years and learning why is like digging deep in someone's backyard. Walls can develop between subject matter experts and the project team when respect is not part of the process.

Not only may the original review be potentially unsettling, but as activities are transitioned to being done electronically, many manual processes are eliminated. It could be that a person who has been doing process steps 1 to 15 for many years now sees that a majority of those steps will go away. This can cause them to begin to question the need for them within the organization. This can be demoralizing and may make them leery of the project. This issue needs to be addressed with them straightforwardly.

As it turned out, there were several opportunities to improve the customer response process and decrease the response rate to be even lower than the original objective.

It was also determined that it was no longer necessary to have three customer service representatives. Only one would be needed to handle the revised workload. In this case, one representative was offered another position that had recently been vacated. The remaining position was terminated.

The process was improved by eliminating wasteful activities. However, it was unfortunate that one representative had to transition to another job, and even more unfortunate that the third representative lost their job.

Tips for Managing Change Successfully

Project managers must not only manage changes to the project, but they must manage "people change" constructively. Here are a few tips that will help make things go smoother.

Respect the people whose lives are being rearranged by the change. One way to do this is to remember how difficult it may have been for you to cope with a change in your life.

Support the people who are challenged by the change during the execution phase. This is when reality sets in as the changes are tested. For the first time, people can see the new way of doing things. The current state is transitioning to the future and the people affected are learning new procedures

Identify project champions among the stakeholders. Reach out to those most likely to benefit from the change and look for individuals are known and respected in the organization Ask these people to champion the new way of doing things by sharing their positive experiences, helpful tips for learning the new way of doing things, and the benefits they expect to gain from the change.

Maintain a positive "can-do" attitude. This can make any transition better just by itself. Find ways to show people how their lives will improve while also listening to their perspectives and working with them to solve what to them may seem like obstacles.

Change Management Documentation

The outputs for change management monitoring and controlling typically include:

- Approved change requests
- Change request log
- Updated project documents

Points of Control

Internal and external changes to a project will occur during the project's execution phase. Experts in project management say there are many points of control in managing a project.

The following 12 possible points of control are often used:

1. Milestone hit rate
2. Deliverable hit rate
3. Variance from the critical path
4. Actual costs versus estimated costs
5. Issues management
6. Number of risk items
7. Team status
8. Sponsor/management/client time commitment
9. Resource shortfall
10. Critical skills shortfall
11. Personnel turnover rate
12. Training completed versus training planned

These are not the only points of control available to you as a project manager. Nor are you expected to use all of them on every project. But project managers must rely on metrics or quantifiable items that can be tracked. This will help you identify project changes and departures from expectations, and this is how control is established and maintained.

What to do:	**Other actions:**
☐ Identify the information needed on your project's change request log.	☐ ______________________
☐ List several ways a "can do" attitude will help you manage organizational change.	☐ ______________________
	☐ ______________________
☐ Review the 12 points of control and identify the best ones for your project.	☐ ______________________
	☐ ______________________

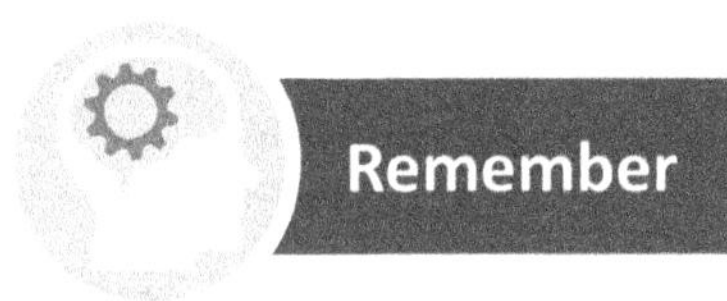

Remember

- ✓ The change request log provides a dashboard look at the state of a project's change requests, depending on how you set up the format.
- ✓ Resistance to change is natural, but to keep the project on track, you must manage the resistance.
- ✓ A change control board is intended to protect a project from inappropriate change requests. It will help you evaluate trade-offs in scope, schedule, and cost.

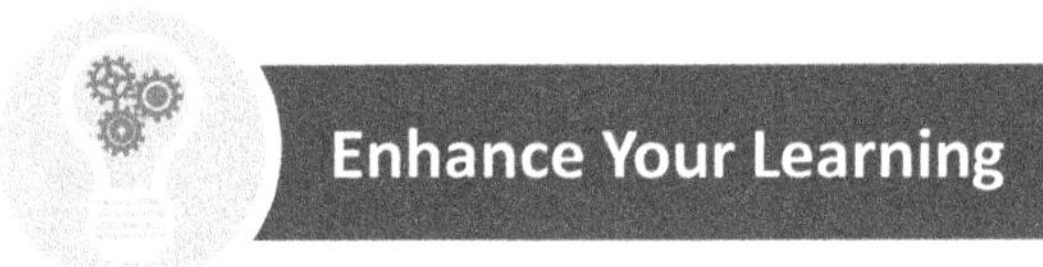

Enhance Your Learning

View the following 5-minute video by strategy+business about change management and learn why only 54% of major change initiatives are successful.

Aguirre, D. (2014). *How to Lead Change Management.*		Available at: https://www.youtube.com/watch?v=PQ0doKfhecQ

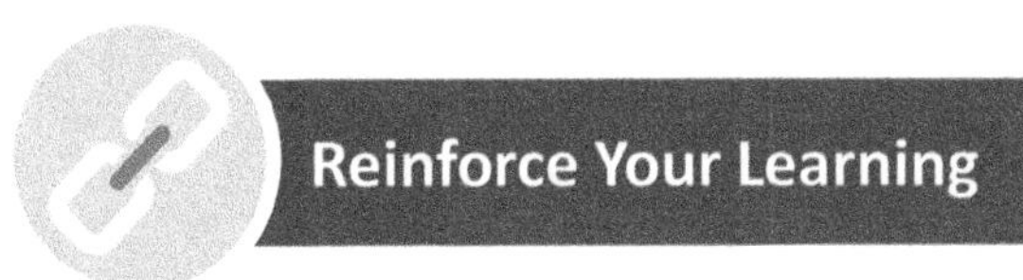

Implementing Changes

1. What is the most important thing you can do to minimize the impact of internal change? External change?

2. Supporting Organizational Change. Read the following scenario, then review the communication questions and follow the instructions for developing your action plan.

Scenario: <u>Major Building Renovation</u>. Assume you have been asked to spearhead a project to relocate (for two years) your business unit's facilities, including operations and administrative space. Responsibilities include integration and coordination of all department activities, as well as selected auxiliary departments housed locally. All implementation efforts must promote a collaborative environment and positive impact.

<u>Communication Questions</u>   <u>Planned Actions</u>
To whom should it be communicated?

How should it be communicated?

What communication methods should be used?

Change before you have to.

—Jack Welch

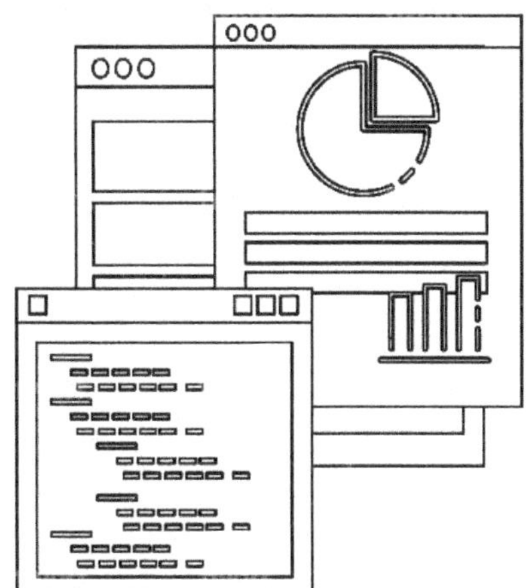

Summary

A Final Word: Monitoring and Controlling Projects

This chapter focused on monitoring and controlling all aspects of a project. We emphasized the types of things that can impact a project during the execution phase, but you can apply the concepts discussed to all phases of the project life cycle.

Information you reviewed in this chapter include:

- Appreciating how scope variation must be managed from all project perspectives
- Recognizing the key components of a Project Management Plan
- Describing all aspects of a project that need monitoring and controlling
- Understanding specific ways monitoring and controlling can be accomplished
- Expressing the advantages of early detection and correction for project metric (scope, schedule, and cost) variations

In the next two *Mastering Project Management* books *Results Through People* and *Controlling Time, Money, and Risk*, we focus on the people and business sides for project management.

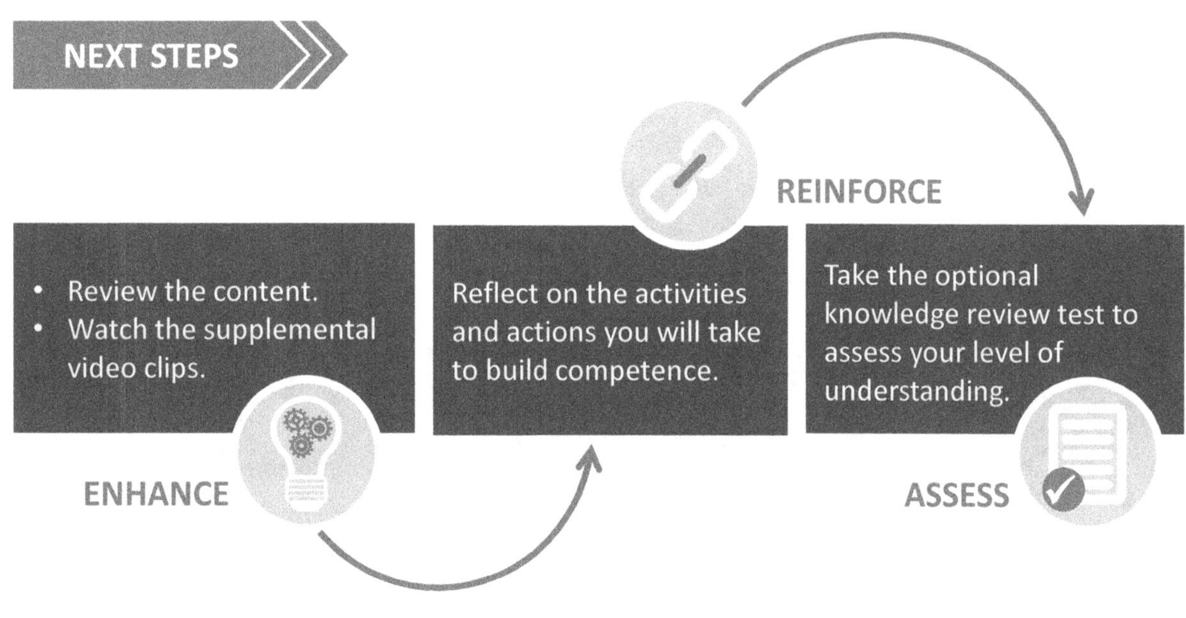

Notes:

Monitoring and Controlling Projects—Recap Checklist

During the execution phase of a project, the project manager must monitor and control the project to keep it on track and resolve potential problems before they impact project success.

1. Differentiate Between Monitoring and Controlling a Project

- ☐ Review the project charter and scope statement for updates and changes.
- ☐ Be mindful that there will be changes to the project's deliverables.
- ☐ Check the project planning for monitoring and controlling guidelines.

2. Start the Execution Phase by Reviewing the Critical First 10 Percent

- ☐ For a project you worked on, did the team follow the project plan at the start of the execution phase?
- ☐ Revisit the tips to execute a successful project. Do you agree that monitoring is management?
- ☐ Do you often ask people to re-do work? If so, what could you do to help avoid that problem?

3. Review the Components of a Project Plan

- ☐ Document and communicate updates to your plan: why, who, what, when, and how.
- ☐ Review your baseline project management plan and set up triggers to help monitor your project to ensure it is going as planned.
- ☐ Control your project by proactively taking corrective measures to get back on track.

4. Monitor and Control Scope Creep

- ☐ Review the requirements for your project and validate them to this point in the project.
- ☐ Review the project change request log to ensure appropriate actions have been taken.
- ☐ Review the scope baseline thresholds to ensure compliance. Create thresholds if they do not exist.

5. Monitor and Control Project Resources

- ☐ Monitor resources to ensure they are allocated according to the project plan. If not, how could the problem be resolved?
- ☐ Consider how much time you spend observing and discussing issues with others.
- ☐ Identify an interpersonal skill that someone you work with does well and compliment the person.

6. Identify Components of Risk Planning and Management

- ☐ Review the issues and change request log examples and consider how risks are tracked.
- ☐ Review the risks in your project and consider what tools you could use to improve risk management.
- ☐ Review your project's monetary reserves and consider recent project performance. Is the reserve likely to be adequate?

7. Review Quality Monitoring and Control Tools

- ☐ Review a cause-and-effect diagram for your project and validate it.
- ☐ Review the quality management plan for specifics of monitoring and controlling quality.
- ☐ Review a process flow for your project for potential improvements.

8. Monitor and Control the Procurement Process

- ☐ Review the contract change control system that applies to your project.
- ☐ Verify the procurement activity that is associated with your responsibilities within the project. Are they on track with the original baseline?
- ☐ Check to see how suppliers are doing on your project's procurement scorecard.

9. Monitor and Control Project Communication

- ☐ Review your most current project status report. Is what you need to know included?
- ☐ Identify the types of communication you are receiving and determine their effectiveness. Can you suggest improvements?
- ☐ Consider the client/stakeholders on your project. Do you think they know the project's current status?

10. Implement Change Management Monitoring and Controlling

- ☐ Identify the information needed on your project's change request log.
- ☐ List several ways a "can-do" attitude will help you manage organizational change.
- ☐ Review the 12 points of control and identify the best ones for your project.

Action Planning | **Competency #1** >>

Resource Management
Demonstrates awareness of technical resources; knows how to apply resources to achieve desired outcomes.

Briefly describe how improvement in this competency will help you achieve important results or better meet your job responsibilities.

List courses, books, and independent study opportunities that could help you develop this competency.

Action Planning | **Competency #1** >>

Identify one or more people who could help you, either as a role model or source of information. Write any questions you want to ask each person.

What specific steps will you take? **Start Date** **Finished**

Action Planning | **Competency #2** >>

Management Controls

Ensures the integrity of the organization's processes; promotes ethical and effective practices.

Briefly describe how improvement in this competency will help you achieve important results or better meet your job responsibilities.

List courses, books, and independent study opportunities that could help you develop this competency.

Identify one or more people who could help you, either as a role model or source of information. Write any questions you want to ask each person.

What specific steps will you take? **Start Date** **Finished**

Action Planning | **Competency #3** >>

Oral and Written Communication

Makes clear and effective presentations to individuals and groups; listens to others; communicates effectively in writing; can critically review and comprehend information written by others.

Briefly describe how improvement in this competency will help you achieve important results or better meet your job responsibilities.

List courses, books, and independent study opportunities that could help you develop this competency.

Action Planning | **Competency #3** >>

Identify one or more people who could help you, either as a role model or source of information. Write any questions you want to ask each person.

What specific steps will you take?   **Start Date**   **Finished**

Failure is simply the opportunity to begin again,
this time more intelligently.

—Henry Ford

Monitoring and Controlling Projects

Knowledge Review Test

Part A. Knowledge Review Test—Questions

Part B. Knowledge Review Test—Answer Sheet

Part A. Knowledge Review Test—Questions

Monitoring and Controlling

1. Controlling occurs:
     A. On a continual basis throughout the project.
     B. At the end of each phase of the project life cycle to assess its success in meeting project objectives.
     C. At the end of each planning process.
     D. On a continual basis throughout the executing process and as needed in the other processes.

2. Monitoring entails:
     A. Doing the work.
     B. Exercising influence over the project, taking action to analyze, evaluate alternatives, and recommend corrective action.
     C. Watching over the project, collecting, reporting, and disseminating project performance data and information.
     D. All the answers are correct.

3. During a team meeting the team determines that the client would benefit from several activities of extra work, and they decided to add them to the project. This is called:
     A. Gold plating
     B. Extra scope
     C. An authorized change request
     D. Superior customer service

4. Any variation from the original baseline scope is typically considered a _____ and must be documented accordingly:
     A. Bad thing.
     B. Variable ingredient.
     C. Change request.
     D. Gold plating element.

5. Which of the following are tips for successfully executing a project:
     A. Plan, and always think ahead.
     B. Standardize.
     C. Get it right the first time.
     D. Monitor and control.
     E. All the answers are correct.

6. The components of a project plan are important in change control because they:
     A. Provide the baseline against which changes are managed.
     B. Are expected to change throughout the project.
     C. Alert the project team to issues that may cause problems in the future.
     D. Provide information on project performance.

7. Uncontrolled changes are often referred to as:
    A. Scope creep.
    B. Scope leap.
    C. Gold plating.
    D. All the answers are correct.

8. A scope change is one that:

    A. Results in a modification of all the project baselines.

    B. Ends with a lessons learned.

    C. Changes the project's agreed-upon scope as defined by its WBS.

    D. Requires adjustments in cost, time, quality, and other objectives.

9. Which of the following is an example of quality assurance?

    A. Inspection

    B. Team training

    C. Pareto charts

    D. Fishbone diagrams

10. Which of the following is not part of a well-designed performance management system?

    A. Clarifying expectations.

    B. Helping team members succeed.

    C. Rewarding performance.

    D. Following up on poor performance.

    E. None of the answers are correct

11. Tools and techniques useful in risk identification include which of the following:

    A. Checklists.

    B. Databases.

    C. Flowcharts.

    D. Interviews.

    E. All the answers are correct.

12. Unplanned responses to negative risk events are often called:

    A. Workarounds.

    B. Risk avoidance.

    C. Workouts

    D. Risk mitigation.

13. Quality assurance is considered synonymous with prevention and quality control is often considered synonymous with inspection. Common tools and techniques for managing quality may include which of the following:

    A. Observation and audits.

    B. Cause and effect diagrams.

    C. Root cause analysis.

    D. Statistical sampling and inspection.

    E. All the answers are correct.

14. Project quality management was once thought to include only inspection or quality control. In recent years, the concept of project quality management has broadened. Which statement is not representative of the new definition of quality management?

    A. Quality is designed into the product or service, not inspected into it.

    B. Quality is the concern of the quality assurance staff.

    C. Customers require a documented and, in some cases, registered quality assurance system.

    D. National and international standards and guidelines for quality assurance systems are available.

15. The idea behind implementing a procurement system is that once implemented, it can streamline the procurement monitoring and control efforts. Such a system would include all the following except:

    A. Payment system.

    B. Expert witness training.

    C. Claims administration.,

    D. Record management system.

16. A change control system should include all the following except:

    A. Paperwork, tracking systems, and approval levels necessary for authorizing changes.

    B. Procedures for conducting a mid-project control system review.

    C. Procedures for automatic approval of defined categories of change.

    D. A description of the powers and responsibilities of the change control board.

17. The outputs for communication monitoring and control typically include all the following except:

    A. Project communications in all forms.

    B. Project charter.

    C. Work performance information.

    D. Change requests.

    E. Updated project documents.

18. An Ishikawa diagram helps:

    A. Explore past outcomes

    B. Stimulate thinking, organize thoughts and generate discussion

    C. Show team responsibilities

    D. Show functional responsibilities

19. The best description of resilience is:

    A. The ability to block out change.

    B. The ability to deal with many changes simultaneously.

    C. The ability to absorb high levels of disruptive change while displaying minimal dysfunctional behavior.

    D. The ability to deal with employees resisting change and effectively handle ongoing activities of the project.

20. Which of the following conflict resolution techniques will generate the most long-lasting solution?

    A. Forcing

    B. Compromising

    C. Smoothing

    D. Problem solving

Part B. Knowledge Review Test—Answer Sheet

1.	_X_ A	___ B	___ C	___ D	___ E
2.	___ A	___ B	_X_ C	___ D	___ E
3.	_X_ A	___ B	___ C	___ D	___ E
4.	___ A	___ B	_X_ C	___ D	___ E
5.	___ A	___ B	___ C	___ D	_X_ E
6.	_X_ A	___ B	___ C	___ D	___ E
7.	___ A	___ B	___ C	_X_ D	___ E
8.	___ A	___ B	_X_ C	___ D	___ E
9.	___ A	_X_ B	___ C	___ D	___ E
10.	___ A	___ B	___ C	___ D	_X_ E
11.	___ A	___ B	___ C	___ D	_X_ E
12.	_X_ A	___ B	___ C	___ D	___ E
13.	___ A	___ B	___ C	___ D	_X_ E
14.	___ A	_X_ B	___ C	___ D	___ E
15.	___ A	_X_ B	___ C	___ D	___ E
16.	___ A	___ B	_X_ C	___ D	___ E
17.	___ A	_X_ B	___ C	___ D	___ E
18.	___ A	_X_ B	___ C	___ D	___ E
19.	___ A	___ B	_X_ C	___ D	___ E
20.	___ A	___ B	___ C	_X_ D	___ E

Afterword

Reading is a valuable way to gain knowledge. This is why schools have textbooks, why online news sites and blogs are popular, and why business and self-help books are a billion-dollar industry. However, reading without reflecting and acting will not be productive.

As with any learning, simply reading without engaging will accomplish little. The level of effort you put forth will determine how much you improve and how close you come to realizing your full potential. You must reflect on what you learn and then apply that knowledge to your projects and career.

How do you do that? You do it by acting; by practicing what you have learned. This Book 1 is designed around the following four themes and chapters:

1. Contemporary Project Management

2. Defining Project Scope

3. Project Planning and Scheduling Tools

4. Monitoring and Controlling Projects

Mastering Project Management is a three-part book series designed for individuals who manage or anticipate managing workplace projects, and this book ***Planning For Performance*** is just one of three in the series. The two companion books, ***Results Through People,*** and ***Controlling Time, Money, and Risk,*** are focused on the people and business sides of project management, respectively.

To become a more accomplished project manager or a more successful leader, you must engage with the material presented here. Question it, think about how it relates to your projects, and begin to internalize the concepts.

As Jim Lovell, Apollo 13 astronaut and famous author once said, "There are people who make things happen, there are people who watch things happen, and there are people who wonder what happened. To be successful, you need to be a person who makes things happen."

Action is the foundational key to all success.

—Pablo Picasso

References

Adams, S. (2013). How to communicate effectively at work. *Forbes.* http://www.forbes.com/sites/susanadams/2013/11/19/how-to-communicate-effectively-at-work-3/

Aguirre, D. (2014). *How to Lead Change Management.* [Video]. YouTube. https://youtu.be/PQ0doKfhecQ

Akbar, (2022). *Project Monitoring vs Controlling, Difference B/W Project Monitoring and Controlling.* [Video]. YouTube. https://www.youtube.com/watch?v=XTo99IsVS14&t=1s

Akinwale, P. (1016). *Manage Communications/Stakeholder vs Control Communications/Stakeholder.* [Video]. YouTube. https://www.youtube.com/watch?v=ZKYmRKadujg

Andrade, R. (2013). *Affinity Diagrams.* [Video]. YouTube. https://www.youtube.com/watch?v=jvTSsJrDZec

Bourne, L. (2009). *Stakeholder Relationship Management: A Maturity Model for Organizational Implementation.* Routledge.

Bridges, J. (2019). *How to Monitor Daily Progress as a Project Manager.* [Video]. YouTube. https://www.youtube.com/watch?v=E2Qahnv1j1g

Bridges, J. (2019). *How to Successfully Execute a Plan.* [Video]. YouTube. https://www.youtube.com/watch?v=802yQd8TNf8&t=1s

Bridges, J. (2017). *Top 5 Excel Project Templates - Project Management Training.* [Video]. YouTube. https://www.youtube.com/watch?v=BAY78mJr9cQ

Bridges, J. (2018). *What is Project Scheduling?* [Video]. YouTube. https://www.youtube.com/watch?v=WNWSQOynrl0

Burke, C. (2021). *How to Create Effective Project Deliverables.* [Video]. YouTube. https://www.youtube.com/watch?v=7HADacRHSFs

Christianson, S. (2019). *Project Management Networks Part 2: Forward and Backward Pass.* [Video]. YouTube. https://www.youtube.com/watch?v=Momnar3Mr9Q

Clayton, M. (2022). *How to Create a Gantt Chart in 9 Easy Steps.* [Video]. YouTube. https://www.youtube.com/watch?v=FX70rWEE8eY

Clayton, M. (2019). *Project Management in Under 5: What is Project Integration?* [Video]. YouTube. https://www.youtube.com/watch?v=cHUP61KU3xE

Clayton, M. (2017). *What is a PERT Chart? Project Management in Under 5.* [Video]. YouTube. https://www.youtube.com/watch?v=i160aaBX7mE

Clayton, M. (2017). *What is Milestone Planning?* [Video]. YouTube. https://www.youtube.com/watch?v=aJcd1rQ0Z9w

Cobb, A. (2011). *Leading Project Teams: The Basics of Project Management and Team Leadership.* SAGE Publications, Inc.

Cooper, R. (2005). *Product Leadership: Pathways to Profitable Innovation*. Basic Books.

Craft, C. (2022). *Project Management QuickStart Guide: The Simplified Beginner's Guide to Precise Planning, Strategic Resource Management, and Delivering World Class Results*. ClydeBank Media LLC.

Creative Education. (2020). *How to draw a Project Network Diagram.* [Video]. YouTube. https://www.youtube.com/watch?v=pWkhR67bkXw

Dartmouth University. (2014). *Five-Level Work Breakdown Structure.* [Video]. YouTube. https://www.youtube.com/watch?v=SqcuLsyFr-o

Deen, D. (2011). *Top 5 Project Tracking & Monitoring Tips for Project Managers*. [Video]. YouTube. https://www.youtube.com/watch?v=CTg_9K5K6hE

Denziel, G. (2004). "E-mail Use and Communication Perceptions of University of Vermont Extension Employees." *Journal of Extension* 42(3).

Dodd, J. (2014). *The Role of The Project Manager*. [Video]. YouTube. https://www.youtube.com/watch?v=dUhJuB69ZBo

Donahue, W. (2022). *Building Leadership Competence: A practical guidebook for professional development*. Second Edition. IngramSpark.

Donahue, W. (2020). *Small Projects Tool - EMPTY SPREADSHEET*. https://app.box.com/s/gq1t096g0cs51bbi4dwzi681xutzrh72

Donahue, W. (2021). *Unlocking Lean Six Sigma*. IngramSpark.

Dow, W. (2008). *Project Management Communications Bible*. Wiley.

Faraaz, A. (2019). *Identify Stakeholders|Project Stakeholder Management.* [Video]. YouTube. https://www.youtube.com/watch?v=0xa2Q2XU2nM

Gido, J., Clements, J. (2017). *Successful Project Management*. Cengage Learning.

Gilbert, T. (1996). *Human Competence: Engineering Worthy Performance.* ISPI.

Girdler, A. (2020). *Project Charter Guide [HOW TO WRITE A PROJECT MANAGEMENT CHARTER]*. [Video]. YouTube. https://www.youtube.com/watch?v=FIPeKqrvNCs

Goulston, M. (2015). *Just Listen: Discover the Secret to Getting Through to Absolutely Anyone*. AMACOM.

Hall, H. (2022). *10 Ways to Engage Project Stakeholders*. https://projectriskcoach.com/engage-project-stakeholders/

Harvey, J. *The Abilene Paradox*. Video Production.

Haworth, S. (2018). *Start Your Projects Right: A Complete Guide to Project Initiation*. Digital Project Manager. https://thedigitalprojectmanager.com/project-initiation-guide-how-start-projects/

Janis. (2020). *What is Level of Effort?* [Video]. YouTube. https://www.youtube.com/watch?v=u5JJOiSlaSI

Kaufman, A. (2017). *How to Calculate Critical Path: Project Management Professional (PMP)® Exam Prep.* [Video]. YouTube. https://www.youtube.com/watch?v=0Lo4zsB-bjE

Kaufman, A. (2010). *Leading cause of project failure: Poor project planning*. [Video]. YouTube. http://www.youtube.com/watch?v=MwQ_5JfDIJs

Kaufman, A. (2015). *Sequence Activities Demonstration*. [Video]. YouTube. https://www.youtube.com/watch?v=H-1Ab30_rrM

Knowledgehut. (2019) *4 Stages of Project Life Cycle | Phases of Project Management Life Cycle*. [Video]. YouTube. https://www.youtube.com/watch?v=N3N9-RLSbvo

Kogon, K. (2015). *Project Management for the Unofficial Project Manager: A FranklinCovey Title*. BenBella Books

Kouzes, J.M. & Posner B.Z. (2017). *The Leadership Challenge*. San Francisco: Jossey-Bass, Inc.

Kumar, K. (2016). *Organization structure Influence project management | Functional Matrix Projectized organization*. [Video]. YouTube. https://www.youtube.com/watch?v=pbtqqZm8Ak0

Moore, S. (2009). *Strategic Project Portfolio Management: Enabling a Productive Organization.* Microsoft Executive Leadership Series.

Mulcahy, R., (2020). *Rita Mulcahy's Hot Topics*. 10th edition. RMC Publications.

Mulcahy,R., (2022). *Rita Mulcahy's PMP Test Prep.* 10th edition. RMC Publications.

Nutcache. (2022). *10 Attributes of an Effective Project Manager*. https://www.nutcache.com/blog/10-attributes-effective-project-manager/

People Rich. (2012). *Project Management 2 – Scope Management*. [Video]. YouTube. https://youtu.be/xjawa7wnd-w

Pritchard, C. (n.d.). *The Project Management Communications Toolkit*. Artech.

ProjectManagement.com. (2021). *Top 10 Reasons Why Projects Fail.* https://project-management.com/top-10-reasons-why-projects-fail/

Project Management Institute (PMI). (2015). *Pulse of the Profession: Capturing the Value of Project Management*. http://www.pmi.org/learning/pulse.aspx

Project Management Institute (PMI). (2017). *A Guide to the Project Management Body of Knowledge (PMBOK® Guide)*-Sixth Edition.

Project Management Institute (PMI). (2021). A Guide to the Project Management Body of Knowledge (PMBOK® Guide)-Seventh Edition.

Project Management Institute (PMI). (2017). *Q&As For The PMBOK® Guide Sixth Edition.*

Pritchard, C. (2013). *The Project Management Communications Toolkit*. Artech.

RiskDoctorVideo. (2015). *The Wasa – a true story of scope creep*. [Video]. YouTube. https://www.youtube.com/watch?v=kmJ59yyYza4

Scott, M. & Brydon, S. (1997). *Dimensions of Communications: An Introduction.* Mountain View: Mayfield Publishing Co.

Simplilearn. (2012). *Monitor and Control Risk|Project Risk Management.* [Video]. YouTube. https://www.youtube.com/watch?v=TdWd_4glh_0

Steps in the communication process. (2013). http://thecommunicationprocess.com/the-communication-process/

The Crowd Training. (2016). *Drawn Out Project Scope Management.* [Video]. YouTube. https://www.youtube.com/watch?v=KZn2ly_p8pY

Thompson, J. (2011). Is nonverbal communication a numbers game? *Psychology Today.* http://www.psychologytoday.com/blog/beyond-words/201109/is-nonverbal-communication-numbers-game

Tuckman, B. (1965). Developmental sequence in small groups. *Psychological Bulletin*, Vol 63(6), Jun 1965, 384-399Whitt, J. (2013). *Conflict Resolution Training: How to Manage Team Conflict in Under 6 Minutes!* [Video]. YouTube. https://youtu.be/PHJ8eybXJdw

Whitt, J. (2013). *Controlling Project Scope: Top 5 Project Scope Control Tips.* [Video]. YouTube. https://www.youtube.com/watch?v=D7Jx1ob1sPg

Whitt, J. (2014). *The project Management Lifecycle.* [Video]. YouTube. https://www.youtube.com/watch?v=POuGKD3xLqs&t=7s

Whitt, J. (2018). *Top 10 Project Management Responsibilities.* [Video]. YouTube. https://www.youtube.com/watch?v=vMWJyELLSQ4

Zink, T. (2017). *Project Management Tips: Working with Tasks and Milestones in Your Project Schedule.* [Video]. YouTube. https://www.youtube.com/watch?v=tP4wIlWjsqU

Index

A

B

C

[Created with **TExtract** /
    www.TExtract.com]

324

*A good plan violently executed now is better than a
perfect plan executed next week.*

—George S. Patton

About the Author

Wesley E. Donahue, PhD, PE, PLS, PMP®, Six Sigma Blackbelt

As a business owner, engineer, manager, and now educator, I have always thought of myself as being in the business of helping other people succeed. But like you, at each step of the way I had to learn, and I had to put that learning into action. This latest book, ***Mastering Project Management: Planning For Performance,*** speaks from the voice of experience. It is book one in the series of three complementary books, and is focused on the technical side of project management. The two companion books, ***Results Through People***, and ***Controlling Time, Money, and Risk***, are focused on the leadership and business management sides of project management, respectively.

As for my background, I am a professor of Organization Development at Penn State University. In this capacity, I am engaged in top-ranked graduate research and programming in learning and performance systems and lead a successful online graduate program in organization development and change. I am also president of Centrestar, Inc. a firm that offers a unique and straightforward approach for professionals to assess their leadership skills, develop personalized roadmaps for success, and access on-demand micro-learning courses and competency-based workbooks available on Amazon. View our resources at: https://www.centrestar.com

Prior to my current position at Penn State, I was Director of Penn State Management Development Programs and Services, where I led and provided education and training services to business and industry clients around the world. Before that, I had years of multi-faceted experience as a project manager, leader, and business owner. I was regional sales vice-president for a mid-sized plastics packaging manufacturer; co-founder and executive vice-president of a manufacturing company; corporate and international manager of technology for a *Fortune 200* multi-national company; and I also co-owned and operated a retail business.

I am a registered professional engineer and land surveyor with an MBA, Six Sigma black belt, certified project management professional (PMP®), author of ***Building Leadership Competence***, ***Unlocking Lean Six Sigma***, **Developing Effective Leadership in Organizations**, as well as a host of workbooks and and other education and training materials.

I would enjoy hearing from you and finding out how this book has helped you achieve your goals. Please contact me at wdonahue@centrestar.com

Best wishes for your continued success.

Knowing is not enough; We must apply.
Willing is not enough; We must do.

—Bruce Lee